Understanding Bioanalytical Chemistry

Understanding Bioanalytical Chemistry

Principles and applications

Victor A. Gault and Neville H. McClenaghan

School of Biomedical Sciences
University of Ulster
Northern Ireland, UK

A John Wiley & Sons, Ltd., Publication

Library of Congress Cataloging-in-Publication Data

Gault, Victor A.
  Understanding Bioanalytical Chemistry : principles and
applications / Victor A. Gault and Neville H. McClenaghan.
      p. ; cm.
  Includes index.
  ISBN 978-0-470-02906-0 – ISBN 978-0-470-02907-7
1. Analytical biochemistry – Textbooks. I. McClenaghan, Neville H. II.
Title.
  [DNLM: 1. Biochemistry. 2. Molecular Biology. QU 4 G271b 2009]
  QP519.7.G38 2009
  572'.36 – dc22

                                        2008022162

ISBN: 978-0-470-02906-0 (HB)
        978-0-470-02907-7 (PB)

A catalogue record for this book is available from the British Library

Typeset in 10.5/13pt Times by Laserwords Private Limited, Chennai, India

First Impression   2009

Contents

Preface

Telling first year life and health science students they have to study chemistry as part of their degree programme is often met with disillusionment or despair. To many the very word chemistry conjures up images of blackboards filled with mind-numbing facts and formulae, seemingly irrelevant to their chosen career paths. This textbook is our response to the very many students who plead with their tutors to 'please teach us what we need to know'. Rather than the simplistic interpretation of this statement as an indirect way of asking tutors to 'please tell us what's on the exam paper' we would see this as a more meaningful and reasonable request.

In recent years we have completely overhauled the way in which we teach bio-analytical chemistry. Taking a 'back to the drawing board' approach, we embraced the challenge of carefully considering the key aspects of chemistry every life and health scientist really needs to know. Our goal was to produce a stand-alone first year undergraduate module comprising a discrete series of lectures and practical classes, using relevant real-life examples to illustrate chemical principles and applications in action. This represented a radical departure from the former module in approach and content, and was extremely well received by students, with a marked improvement in student feedback and academic performance.

On reflection we are at a loss as to why it is tradition for life and health science students not to be introduced to the bioanalytical tools of their trade from the outset of their course. To us this is like teaching students the principles of computer science without actually introducing them to a computer and what it can do. With this in mind, we purposely chose to take an applied approach to chemistry, with an introduction to relevant methods and technologies up front, in order to familiarize students with these tools before they encounter and study them in more detail later in their courses.

Our message to students: To argue that life and health scientists don't need chemistry is like arguing that the world is flat. That is, as much as you might be convinced that it is the case, it does not mean that you are correct. Whether we like it or not, the fact is chemistry lies at the heart of the vast majority of scientific disciplines. Given this, it is pretty much impossible to expect that you will really grasp the fundamentals of core disciplines such as physiology, pathophysiology and pharmacology or be prepared for the diverse range of careers in the life and health sciences without at least a basic knowledge of core chemical principles and applications. This book is designed to complement delivery of first year chemistry, focusing on bioanalytical techniques and their real world applications.

Our message to tutors: We know, we've been there; despite all your best efforts, enthusing life and health science students to study (never mind enjoy) chemistry is like trying to encourage a physicist to build a time machine. The task has not been made any easier by the stereotypical stodginess of chemistry, the expansive nature of the subject, or the encyclopaedic nature of the average chemistry textbook. To compound the problem, few academics in life and health science departments either choose or wish to teach chemistry. Often considered the 'poisoned chalice' and the fate of many an unsuspecting fresh-faced newcomer, effective teaching and learning of first year chemistry represents a considerable challenge.

We hope that you will find this book a useful approach to the subject of bio-analytical chemistry and that it will help raise awareness of the vast scope and topics encompassed in what is a rapidly expanding and advancing field. More-over, we hope that studying the content of this book will provide a fundamental introduction to the tools adopted by life and health scientists in the evolving and exciting new age of 'omics', with the promise of personalized medicine and novel approaches to the screening, diagnosis, treatment, cure and prevention of disease.

1 Introduction to biomolecules

Bioanalytical chemistry relies on the identification and characterization of *particles* and *compounds*, particularly those involved with life and health processes. Living matter comprises certain key *elements*, and in mammals the most abundant of these, representing around 97% of dry weight of humans, are: carbon (C), nitrogen (N), oxygen (O), hydrogen (H), calcium (Ca), phosphorus (P) and sulfur (S). However, other elements such as sodium (Na), potassium (K), magnesium (Mg) and chlorine (Cl), although less abundant, nevertheless play a very significant role in organ function. In addition, miniscule amounts of so-called *trace elements*, including iron (Fe), play vital roles, regulating biochemical pathways and biological function. By definition, *biomolecules* are naturally occurring chemical compounds found in living organisms that are constructed from various combinations of key chemical elements. Not surprisingly there are fundamental similarities in the way organisms use such biomolecules to perform diverse tasks such as propagating the species and genetic information, and maintaining energy production and utilization. From this it is evident that much can be learned about the functionality of life processes in higher mammals through the study of micro-organisms and single cells. Indeed, the study of yeast and bacteria allowed genetic mapping before the Human Genome Project. This chapter provides an introduction to significant biomolecules of importance in the life and health sciences, covering their major properties and basic characteristics.

Learning Objectives

- To be aware of important chemical and physical characteristics of biomolecules and their components.

Understanding Bioanalytical Chemistry: Principles and applications   Victor A. Gault and Neville H. McClenaghan
© 2009 John Wiley & Sons, Ltd

- To recognize different classifications of biomolecules.
- To understand and be able to demonstrate knowledge of key features and characteristics of major biomolecules.
- To identify and relate structure–function relationships of biomolecules.
- To illustrate and exemplify the impact of biomolecules in nature and science.

1.1 Overview of chemical and physical attributes of biomolecules

Atoms and elements

Chemical elements are constructed from *atoms*, which are small particles or units that retain the chemical properties of that particular element. Atoms comprise a number of different *sub-atomic particles*, primarily *electrons*, *protons* and *neutrons*. The *nucleus* of an atom contains positively charged protons and uncharged neutrons, and a cloud of negatively charged electrons surrounds this region. Electrons are particularly interesting as they allow atoms to interact (in bonding), and elements to become ions (through loss or gain of electrons). Further topics in atomic theory relevant to bioanalysis will be discussed throughout this book, and an overview of *atomic bonding* is given below.

Bonding

The physical processes underlying attractive interactions between atoms, elements and molecules are termed *chemical bonding*. Strong chemical bonds are associated with the sharing or transfer of electrons between bonding atoms, and such bonds hold biomolecules together. *Bond strength* depends on certain factors, and so-called *covalent bonds* and *ionic bonds* are generally categorized as 'strong bonds', while *hydrogen bonds* and *van der Waal's forces of attraction* within molecules are examples of 'weak bonds'. These terms are, however, quite subjective, as the strongest 'weak bonds' may well be stronger than the weakest 'strong bonds'. Chemical bonds also help dictate the structure of matter. In essence, covalent bonding (electron sharing) relies on the fact that opposite forces attract, and negatively charged electrons *orbiting* one atomic nucleus may be attracted to the positively charged nucleus of a neighbouring atom. Ionic bonding involves *electrostatic attraction* between two neighbouring atoms, where one positively charged nucleus 'forces' the other to become negatively charged (through electron transfer) and, as opposites attract, they bond. Historically, bonding was first considered in the twelfth century, and in the eighteenth century English all-round scientist, Isaac Newton, proposed that a 'force' attached atoms. All bonds can be explained by quantum theory (in very large textbooks), encompassing the *octet rule* (where eight is the magic number when so-called *valence electrons* combine), the *valence shell electron pair repulsion theory* (where valence electrons repel each other in such a way as to determine geometrical shape), *valence bond theory* (including orbital hybridization and resonance) and *molecular orbital theory* (as electrons are found in discrete orbitals, the position of an electron will dictate whether or not, and how, it will participate in bonding). When considering bonding, some important terms are *bond length* (separation distance where molecule is most stable), *bond energy* (energy dependent on separation distance), *non-bonding electrons* (valence electrons that do not participate in bonding), *electronegativity* (measure of attraction of bound electrons in polar bonds, where the greater the difference in electronegativity, the more polar the bond). *Electron-dot structures* or *Lewis structures* (named after American chemist Gilbert N. Lewis) are helpful ways of conceptualizing simple atomic bonding involving electrons on outer valence shells (see Figure 1.1).

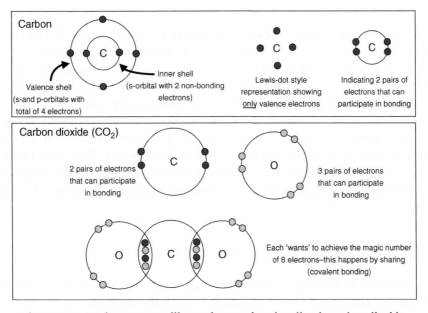

Figure 1.1   **Lewis structures illustrating covalent bonding in carbon dioxide.**

Phases of matter

Matter is loosely defined as anything having mass and taking up space, and is the basic building block of everything. There are three basic *phases of matter*, namely *gas*, *liquid* and *solid*, with different physical and chemical properties. Matter is maintained in these phases by pressure and temperature, and as conditions change matter can change from one phase to another, for example, solid ice converts to liquid water with rise in temperature. These changes are referred to as *phase transitions* inherently requiring energy, following the *Laws of Thermodynamics*. When referring to matter, the word *states* is sometimes used interchangeably with that of *phases*, which can cause confusion as, for example, gases may be in different thermodynamic states but the same state of matter. This has led to a decrease in the popularity of the traditional term *state of matter*. While the general term *thermodynamics* refers to the effects of heat, pressure and volume on physical systems, chemical thermodynamics studies the relationship of heat to chemical reactions or physical state following the basic Laws of Thermodynamics. Importantly, as energy can neither be created nor destroyed, but rather exchanged or emitted (for example as heat) or stored (for example in chemical bonds), this helps define the physical state of matter.

Physical and chemical properties

Matter comprising biomolecules has distinct physical and chemical properties, which can be measured or observed. However, it is important to note that physical properties are distinct from chemical properties. Whereas physical properties can be directly observed without the need for a change in the chemical composition, the study of chemical properties actually requires a change in chemical composition, which results from so-called chemical reactions. Chemical reactions encompass processes that involve the rearrangement, removal, replacement or addition of atoms to produce a new substance(s). Properties of matter may be dependent (*extensive*) or independent (*intensive*) on the quantity of a substance, for example mass and volume are extensive properties of a substance.

Studying physical and chemical properties of biomolecules

A diverse range of bioanalytical techniques have been used to study the basic composition and characteristics of biomolecules. Typically these techniques focus on measures of distinct physical and/or chemical attributes, to identify and determine the presence of different biomolecules in biological samples. This has been important from a diagnostic and scientific standpoint, and some of the major technologies are described in this book. Examples of physical and chemical properties and primary methods used to study that particular property are as follows:

Physical properties: Charge (see ion-exchange chromatography; Chapter 7); Density (see centrifugation; Chapter 6); Mass (see mass spectrometry; Chapter 9); and Shape (see spectroscopy; Chapter 5).

Chemical properties: Bonding (see spectroscopy and electrophoresis; Chapters 5 and 8); Solubility (see precipitation and chromatography; Chapters 6 and 7); Structure (see spectroscopy; Chapter 5).

1.2 Classification of biomolecules

It is important to note that whilst biomolecules are also referred to by more generic terms such as molecules, chemical compounds, substances, and the like, not all molecules, chemical compounds and substances are actually biomolecules. As noted earlier, the term *biomolecule* is used exclusively to describe naturally occurring chemical compounds found in living organisms, virtually all of which contain carbon. The study of carbon-containing molecules is a specific discipline within chemistry called *organic chemistry*. Organic chemistry involves the study of attributes and reactions of chemical compounds that primarily consist of carbon and hydrogen, but may also contain other chemical elements. Importantly, the field of organic chemistry emerged with the misconception by nineteenth century chemists that all organic molecules were related to life processes and that a 'vital force' was necessary to make such molecules. This archaic way of thinking was blown out of the water when organic molecules such as soaps (Michel Chevreul, 1816) and urea (Friedrich Wöhler, 1828) were created in the laboratory without this magical 'vital force'. However, despite being one of the greatest thinkers in the field of chemistry, the German chemist Wöhler was pretty smart not to make too much out of his work, even though it obviously obliterated the vital force concept and the doctrine of vitalism. So from this it is important to remember that not all organic molecules are biomolecules.

Life processes also depend on *inorganic molecules*, and a classic example includes the so-called 'transition metals', key to the function of many molecules (e.g. enzymes). As such, when considering biomolecules it is imperative to understand fundamental features of transition metals and their interaction with biomolecules. Indeed, transition metal chemistry is an effective means of learning basic aspects of inorganic chemistry, its interface with organic chemistry, and how these two fields of study impact on health and disease, and a whole chapter of this book is devoted to this important subject (Chapter 3). There are very many ways of classifying molecules and biomolecules, which often causes some confusion. The simplest division of biomolecules is on the basis of their size, that is, small (*micromolecules*) or large (*macromolecules*). However, while the umbrella term *macromolecule* is widely used, smaller molecules are most often referred to by their actual names (e.g. amino acid) or the more popular term *small molecule*. Yet even the subjective term *macromolecule* and its use are very confused. Historically, this term was coined in the early 1900s by the German chemist Hermann Staudinger, who in 1953 was awarded a Nobel Prize in Chemistry for his work on the characterization of polymers. Given this, the word macromolecule is often used interchangeably with the word *polymer* (or polymer molecule). For the purposes of this book the authors will use the following three categories to classify biomolecules:

Small molecules: The term *small molecule* refers to a diverse range of substances including: lipids and derivatives; vitamins; hormones and neurotransmitters; and carbohydrates.

Monomers: The term *monomer* refers to compounds which act as building blocks to construct larger molecules called *polymers* and includes: amino acids; nucleotides; and monosaccharides.

Polymers: Constructed of repeating linked structural units or monomers, polymers (derived from the Greek words *polys* meaning many and *meros* meaning parts) include: peptides/oligopeptides/polypetides/proteins; nucleic acids; and oligosaccharides/polysaccharides.

1.3   Features and characteristics of major biomolecules

Differences in the properties of biomolecules are dictated by their components, design and construction, giving the inherent key features and characteristics of each biomolecule that enable its specific function(s). There are a number of classes of more abundant biomolecules that participate in life processes and are the subject of study by bioanalytical chemists using a plethora of fundamental and state-of-the-art technologies in order to increase knowledge and understanding at the forefront of life and health sciences. Before considering important biomolecules it is first necessary to examine their key components and construction.

Building biomolecules

Biomolecules primarily consist of carbon (C) and hydrogen (H) as well as oxygen (O), nitrogen (N), phosphorus (P) and sulfur (S), but also have other chemical components (including trace elements such as iron). For now, focus will be placed on the core components carbon, hydrogen, and oxygen, and simple combinations (see also Table 1.1).

Carbon: The basis of the chemistry of all life centres on carbon and carbon-containing biomolecules, and it is the same carbon that comprises coal and diamonds that forms the basis of amino acids and other biomolecules. In other words, carbon is carbon is carbon, irrespective of the product material, which may be hard (diamond) or soft (graphite). Carbon is a versatile constituent with a great affinity for bonding other atoms through *single bonds* or *multiple bonds*, adding to complexity and forming around 10 million different compounds (Figure 1.2). As chemical elements very rarely convert into other elements, the amount of carbon on Earth remains almost totally constant, and thus life processes that use carbon must obtain it somewhere and get rid of it somehow. The flow of carbon in the environment is termed the *carbon cycle*, and the most simple relevant example lies in the fact that plants utilize (or recycle) the gas carbon dioxide (CO_2), in a process called *carbon respiration*, to grow and develop. These plants may then be consumed by humans and with digestion and other processes there is the ultimate generation of CO_2, some of which is exhaled and available again for plants to take up, and so the cycle continues. Being crude, in essence humans and other animals act as vehicles for carbon cycling, being *d*esigned for life in the womb, *d*evouring food and fluids, *d*eveloping, *d*efecating, *d*ying and *d*ecaying, the '6 D's of life'.

Hydrogen: This is the most abundant (and lightest) chemical element, which naturally forms a highly flammable, odourless and colourless diatomic gas (H_2). The Swiss scientist Paracelsus, who pioneered the use of chemicals and minerals in medical practice, is the first credited with making hydrogen gas by mixing metals with strong acids. At the time Paracelsus didn't know this gas was a new chemical element, an intuition attributed to British scientist Henry Cavendish, who described hydrogen gas in 1766 as 'inflammable air', later named by French nobleman and aspiring scientist, Antoine-Laurent Lavoisier, who co-discovered, recognized and named hydrogen (and oxygen), and invented the first *Periodic Table*.

Gaseous hydrogen can be burned (producing by-product water) and thus historically was used as a fuel. For obvious safety reasons helium (He), rather than

Table 1.1   **Examples of simple combinations of carbon, hydrogen and oxygen**

Compound	Chemical formula	Notes
Acetaldehyde	CH_3CHO (MeCHO)	Flammable liquid, fruity smell, found in ripe fruit, and metabolic product of plant metabolism. Chemical associated with the 'hangover' following overindulgence in alcohol.
Acetic acid	CH_3COOH	Hygroscopic liquid, gives vinegar its characteristic taste and smell. Corrosive weak acid.
Acetylene (ethyne)	C_2H_2	Gas containing C to C triple bond. Unsaturated chemical compound which can volatilize carbon in radiocarbon dating.
Benzyl acetate	$C_6H_5CH_2OCOCH_3$	Solid, sweet smelling ester, found naturally in many flowers (e.g. jasmine). Used in perfumes, cosmetics and flavourings.
Carbon dioxide	CO_2	Colourless, odourless and potentially toxic gas which can also exist in solid state (dry ice). Important component of the carbon cycle, a 'greenhouse gas', and contributes to the 'carbon footprint'.
Carbon monoxide	CO	Colourless, odourless and extremely toxic gas, produced by incomplete combustion of carbon-containing compounds (e.g. in internal combustion engines–exhaust fumes).
Ethanol	C_2H_5OH	Flammable, colourless, slightly toxic liquid, found in alcoholic beverages.
Methane	CH_4	Simplest alkane. Gaseous and principal component of natural gas. When burned in O_2 produces CO_2 and H_2O.
Water	H_2O	Normally odourless, colourless and tasteless liquid, but can also exist in solid (ice) or gas (water vapour) states. Non-inert common universal solvent.

hydrogen, was the gas of choice for floatation of Zeppelin airships. Indeed, the now famous Zeppelin airship 'The Hindenburg' was to be filled with He, but because of a US military embargo, the Germans modified the design of the airship to use flammable H_2 gas; an accident waiting to happen, and the rest is history.

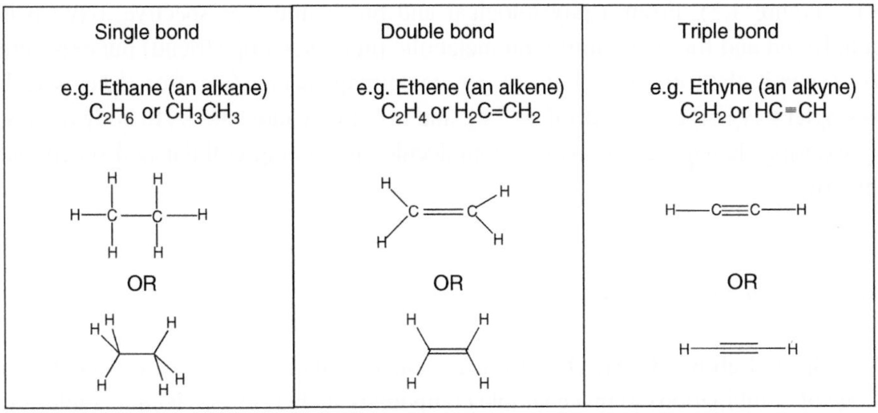

Single bond	Double bond	Triple bond
e.g. Ethane (an alkane) C_2H_6 or CH_3CH_3	e.g. Ethene (an alkene) C_2H_4 or $H_2C=CH_2$	e.g. Ethyne (an alkyne) C_2H_2 or $HC\equiv CH$

Figure 1.2 **Illustration of carbon single, double and triple bonds.**

In terms of biomolecules, hydrogen atoms usually outnumber both carbon and oxygen atoms.

Oxygen: As Lavoisier first generated oxygen from acidic reactions, he falsely believed that it was a component of all acids, deriving the name from the Greek words *oxys* (acid) and *genēs* (forming). Oxygen is usually bonded covalently or ionically to other elements such as carbon and hydrogen, and dioxygen gas (O_2) is a major component of air. Plants produce O_2 during the process of photosynthesis, and all species relying on *aerobic respiration* inherently depend on it for survival. Oxygen also forms a triatomic form (O_3) called *ozone* in the upper layers of the Earth's atmosphere, famously shielding us from UV radiation emitted from the

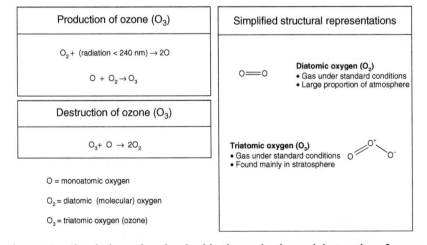

Figure 1.3 **Chemical reactions involved in the production and destruction of ozone.**

Sun (Figure 1.3). From a physiological and biochemical perspective, oxygen is both friend and foe; without it vital metabolic processes stop (friend) but exposure to oxygen in the form of certain oxygen-containing species (e.g. free radicals such as singlet oxygen) can be harmful (foe), and in extreme cases toxic, to body tissues, by exerting damaging actions on biomolecules regulating cellular and functional integrity.

Constructing complex biomolecules

As indicated above, C, H, O and other elements (such as N or P) can bind in a range of combinations to make simple compounds such as those given in Table 1.1. However, the same elements can also bind together to form much more complex structural and functional compounds (or biomolecules) which play vital roles in physiological processes. Major classes of these complex biomolecules are outlined in the boxes below.

Nucleotides

- Nucleotides consist of three components: a heterocyclic nitrogenous base, a sugar, and one or more phosphate groups.

- Nitrogenous bases of nucleotides are derivatives of either purine (adenine, A; or guanine, G) or pyrimidine (cytosine, C; thymine, T; or uracil, U) (see Figure 1.4).

- Nucleotides may be termed *ribonucleotides* (where component sugar is ribose) or *deoxyribonucleotides* (where component sugar is 2-deoxyribose).

- The bases bind to the sugar through glycosidic linkages.

- Also, one or more phosphate groups can bind to either the third carbon (C3) of the sugar of the nucleotide (so-called 3' end) or the fifth carbon (C5) of the sugar (so-called 5' end).

- Nucleotides are structural units of deoxyribonucleic acid (DNA), ribonucleic acid (RNA) and cofactors such as coenzyme A (CoA), flavin adenine dinucleotide (FAD), nicotinamide adenine dinucleotide (NAD) and nicotinamide adenine dinucleotide phosphate (NADP), with important roles in energy transfer, metabolism and intracellular signalling.

- Notably, polynucleotides are acidic at physiological pH due to the phosphate group (PO_4^-); this negatively charged anion is important for bioanalysis.

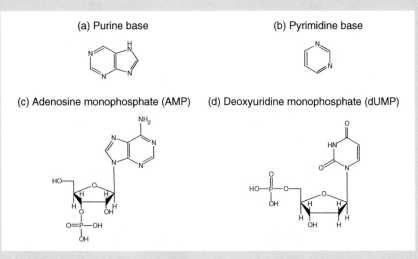

Figure 1.4 Diagrammatic representations of (a) a purine base, (b) a pyrimidine base, (c) a ribonucleotide, adenosine monophosphate (AMP) and (d) a deoxyribonucleotide, deoxyuridine monophosphate (dUMP).

Nucleic acids (e.g. RNA and DNA)

- Nucleic acids are polymers constructed from nucleotides (monomers) and found in cell nuclei.

- RNA comprises ribonucleotides while DNA contains deoxyribonucleotides.

- RNA can comprise the bases adenine (A), cytosine (C), guanine (G), and uracil (U).

- DNA can comprise the bases adenine (A), cytosine (C), guanine (G), and thymine (T).

- A nucleotide comprising a nucleic acid joins with another nucleotide through a so-called phosphodiester bond.

- Polymers of nucleic acids typically have different properties from individual units (nucleic acid monomers).

- There are also structural differences; RNA is usually single-stranded (alpha helix) while DNA is usually double-stranded (double helix). (Figure 1.5)

- DNA can replicate by separation of the two strands of the helix, which act as a template for synthesis of complementary strands.

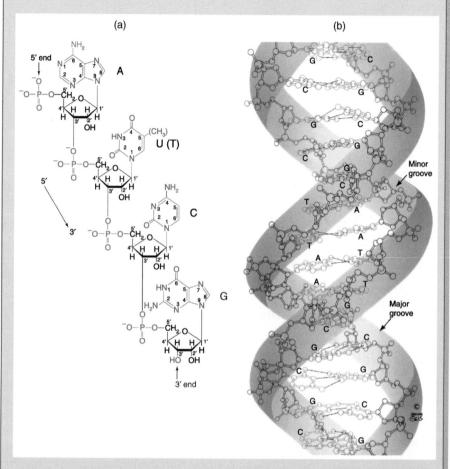

Figure 1.5 Diagrammatic representation of (a) a nucleic acid and (b) double helix structure of DNA. Illustrations, Irving Geiss. Rights owned by Howard Hughes Medical Institute. Reproduction by permission only.

Amino acids

- Molecules that contain a central carbon atom (alpha-carbon) attached to a carboxyl group (COOH), an amine group (NH$_2$), hydrogen atom (H), and a side chain (R group). (Figure 1.6)

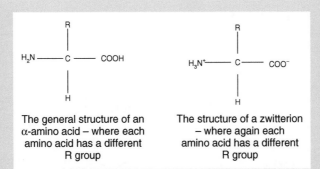

Figure 1.6   General representative chemical structure of an amino acid.

- The R group essentially defines the structure and function of an amino acid; these can generally be classified under three main groups: non-polar, uncharged polar, or charged polar amino acids.

- Amino acids (exceptions include Gly and Cys) are so-called chiral molecules (four different groups attached to alpha-carbon), which means they can exist as two different optical isomers called D (e.g. D-Ala) or more abundant L (e.g. L-Ala).

- There are 20 standard proteinogenic amino acids, of which 10 are essential amino acids that cannot be synthesized in the body so must be derived from the diet.

- Essential amino acids include: Iso, Leu, Lys, Met, Phe, Thr, Trp, Val, Arg and His, where the last two, Arg and His, are only actually essential under certain conditions.

- Amino (NH_2) and carboxylic acid (COOH) groups of the amino acid can readily ionize (to NH_3^+ and COO^-) at certain pHs to form an acid or base.

- The pH at which an amino acid is not in its ionized form (i.e. bears no electric charge) is known as its *isoelectric point*.

- When amino acids contain both positive and negative charges and are electrically neutral they fulfil the criteria of being a zwitterion (dipolar ion), which are highly water-soluble.

- Amino acids can be polymerized to form chains through condensation reactions, joining together by so-called peptide bonds, and they are often referred to as the building blocks of peptides and proteins.

Table 1.2   Classification of essential amino acids

Classification	Name	3-letter code	1-letter code	Structure
Non-polar	Glycine	Gly	G	
	Alanine	Ala	A	
	Valine	Val	V	
	Leucine	Leu	L	
	Isoleucine	Ile	I	
	Methionine	Met	M	
	Proline	Pro	P	
	Phenylalanine	Phe	F	

Table 1.2 (*continued*)

Classification	Name	3-letter code	1-letter code	Structure
	Tryptophan	Trp	W	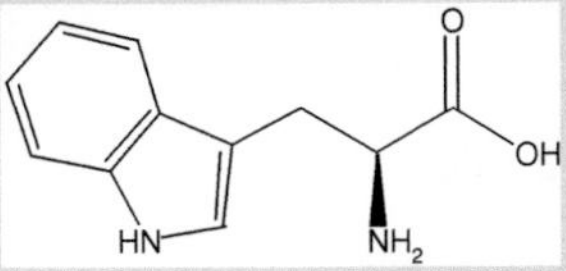
Uncharged polar	Serine	Ser	S	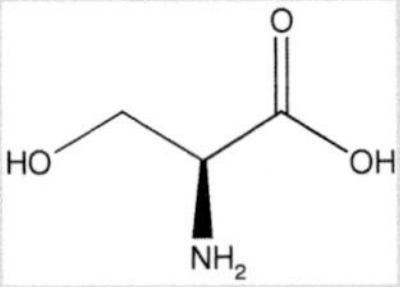
	Threonine	Thr	T	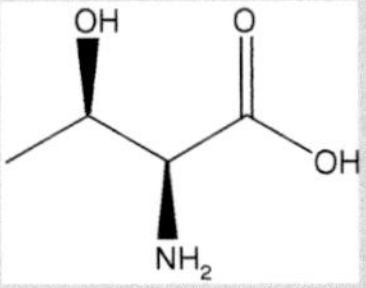
	Asparagine	Asn	N	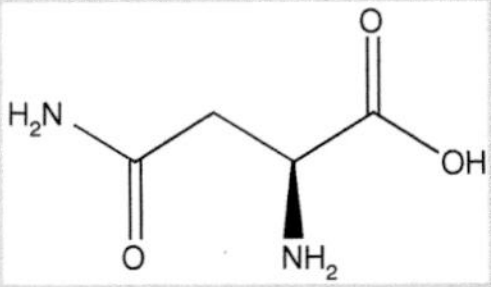
	Glutamine	Gln	Q	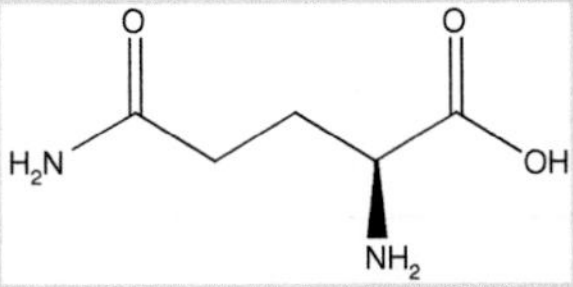
	Tyrosine	Tyr	Y	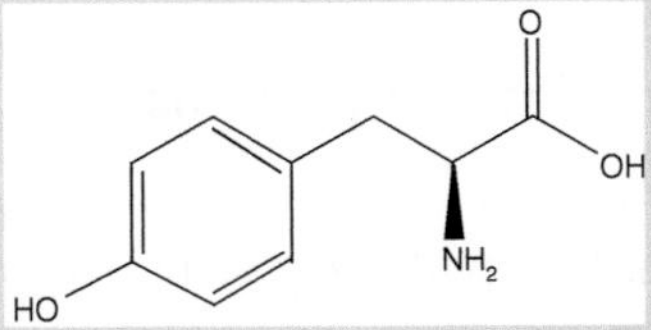
	Cysteine	Cys	C	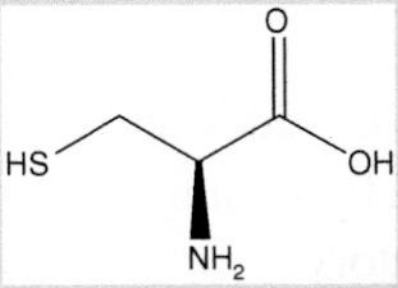

(*continued overleaf*)

Table 1.2   (*continued*)

Classification	Name	3-letter code	1-letter code	Structure
Charged polar	Lysine	Lys	K	
	Arginine	Arg	R	
	Histidine	His	H	
	Aspartic acid	Asp	D	
	Glutamic acid	Glu	E	

Peptides and proteins

- Each peptide or protein is constructed as a string or chain of amino acids, creating huge numbers of variants, analogous to how letters of the alphabet make words.

- Importantly, peptides and proteins are assembled from amino acids on the basis of genetic coding (gene-nucleotide sequence), or can be synthesized in the laboratory.

- Peptides and proteins have individual and unique amino acid sequences (residues), giving them unique structural conformations and biological activity.

- A peptide is a short molecule formed by amino acids linked through amide (peptide) bonds. (Figure 1.7)

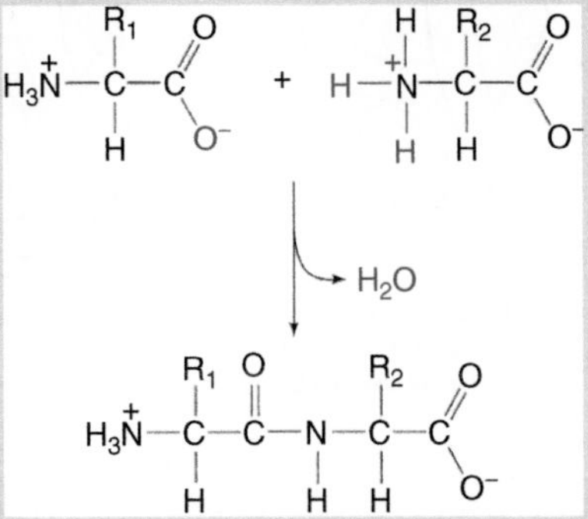

Figure 1.7   Formation of a peptide (amide) bond. From Voet, Voet & Pratt Fundamentals of Biochemistry, 2nd edn; © 2006 Voet, Voet & Pratt; reprinted with permission of John Wiley & Sons, Inc.

- When two amino acids join they form a dipeptide, three a tripeptide and so forth.

- Oligopeptides usually comprise between 3 and 10 amino acids and peptides with more than 10 are often referred to as *polypeptides*.

- Some peptides are called *peptide hormones* and as the name suggests these peptides have hormonal (endocrine) function.

- Proteins comprise one or more polypeptides and while peptides are short, polypeptide proteins are long.

- There are different ways of classifying proteins and one of the most important is on the basis of their structure.

- Biochemically there are four major classifications of protein structure: primary (amino acid sequence), secondary (local spatial arrangement), tertiary (overall 3D structure) and quaternary (protein complex structure). (Figure 1.8)

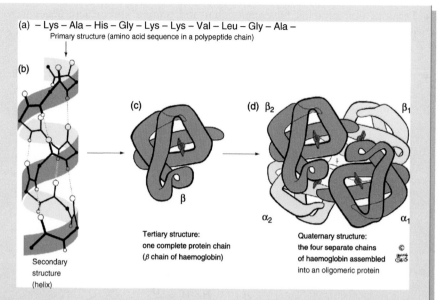

(a) – Lys – Ala – His – Gly – Lys – Lys – Val – Leu – Gly – Ala –
Primary structure (amino acid sequence in a polypeptide chain)

(b)

(c)

(d) β$_2$ β$_1$

β

α$_2$ α$_1$

Secondary
structure
(helix)

Tertiary structure:
one complete protein chain
(β chain of haemoglobin)

Quaternary structure:
the four separate chains
of haemoglobin assembled
into an oligomeric protein

Figure 1.8   Diagrammatic representations of secondary, tertiary and quaternary protein structures. From Voet, Voet & Pratt Fundamentals of Biochemistry, 2nd edn; © 2006 Voet, Voet & Pratt; reprinted with permission of John Wiley & Sons, Inc.

Carbohydrates

- Simple, neutral biomolecules composed of C, H and O, often referred to as *saccharides*.

- All carbohydrates have an aldehyde (aldose) or ketone (ketose) functional group (containing C=O) and a hydroxyl group (−OH).

- Importantly, as carbohydrates contain aldehyde or ketone groups they undergo the same reactions as individual aldehyde or ketone molecules (e.g. oxidation and reduction reactions).

- Classification is based on the number of structural sugar units (and aldehyde or ketone group) in the chain, where 1 unit makes a monosaccharide (e.g. glucose), 2 units are disaccharides (e.g. lactose), 3–10 units are oligosaccharides (e.g. raffinose) and greater than 10 sugar units are polysaccharides (e.g. starch). (Figure 1.9)

- Sugar units are joined together through oxygen atoms, forming a so-called glycosidic bond.

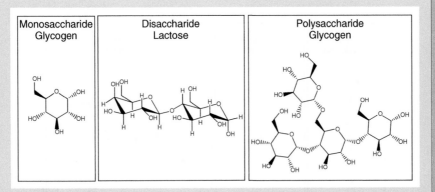

Figure 1.9   Examples of a monosaccharide, disaccharide and polysaccharide.

- Polysaccharides can reach many thousands of units in length, and carbohydrates can contain one or more modified monosaccharide units adding to complexity.

- Carbohydrates are abundant biomolecules in plants (produced by photosynthesis) and animals, with important roles in energy storage, and transport and structural components.

- Nutritionally, carbohydrates may generally be classified as simple sugars (e.g. glucose or fructose), polysaccharides (e.g. homoglycans such as starch or beta-glucans) or complex carbohydrates (e.g. glycoproteins).

Lipids

- Lipids have polar heads (are water-loving, hydrophilic) and long hydrocarbon tails (which are water-repelling, hydrophobic).

- Unlike other biomolecules, lipids are generally not made from repeating monomeric units.

- While the term *lipid* is often used interchangeably with the word fat, fat is actually a subgroup of lipids called *triglycerides* (triacylglycerols).

- Lipids are generally classified as: fatty acids (e.g. stearic acid), triglycerides (e.g. glycerol), phosphoglycerides (e.g. phosphatidylinositol), sphingolipids (e.g. ceramide), and steroids (e.g. cholesterol). (Figure 1.10)

- Lipids have important roles in energy storage, cell membrane structure and cell signalling.

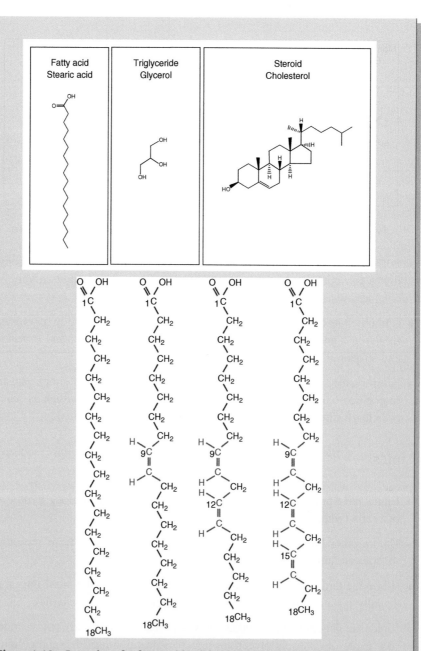

Figure 1.10   Examples of a fatty acid, triglyceride and steroid. From Voet, Voet & Pratt Fundamentals of Biochemistry, 2nd edn; © 2006 Voet, Voet & Pratt; reprinted with permission of John Wiley & Sons, Inc.

1.4   Structure–function relationships

The structure of a given biomolecule will confer certain functional attributes and, as such, is a key defining feature of that biomolecule. There are very many environmental influences that can impact on structure and/or function and, importantly, the stability of the biomolecule. Furthermore, this information is also important when considering bioanalysis, as the outcome of an analytical process or procedure, and indeed the stability of a biomolecule, is dependent on numerous physical and chemical factors including pH, temperature and solvent concentration/polarity. Changes in any or all of these parameters may result in general or specific structural and/or functional changes to a biomolecule that may be *reversible* (e.g. partial unfolding of a protein) or *irreversible* (protein denaturation and degradation). Simple visual examples include the reversible change to hair when it is straightened or curled ('hair perm') and irreversible protein denaturation of 'egg whites' (essentially egg albumins in water) that turn from a transparent liquid into an opaque white solid on cooking (temperature). On a more scientific vein, variations in pH can alter the ionization states of biomolecules such as amino acids in proteins, and phosphate groups in nucleotides. Alterations to *functional groups* can greatly alter the activity and properties of biomolecules, and this is why most physiological, biochemical and enzymatic processes require *homeostatic* conditions (i.e. maintenance of a relatively constant internal environment).

It is also important to map the chemical structure of biomolecules with certain well-defined processes, which may relate to biological activity or chemical reactivity. Medicinal chemists place particular importance on understanding *structure–activity relationships* (SARs) of biomolecules in order to bioengineer modified biomolecules with enhanced activity (*potency*), for example by changing amino acid composition of peptides or insertion/addition of chemical groups. This approach has given new and exciting insights into chemical groups that affect biological processes, and allowed complex mathematical modelling of *quantitative structure–function relationships* (QSARs). This has inherent difficulties, as certain features such as *post-translational modification* of proteins may depend on multiple factors, and thus not all related biomolecules have similar activities (so-called SAR paradox). Historically, one of the first simple examples of QSAR was to predict boiling points on the basis of the number of carbon atoms in organic compounds; more modern applications of QSAR are in drug design and discovery, discussed in more detail in Chapter 12.

1.5   Significance of biomolecules in nature and science

Biomolecules are the essence and currency of life and health processes lying at the heart of the simplest to the most complex system. Understanding the fundamental nature of biomolecules, their structure, location, behaviour and function, is critical to knowledge and understanding of health, the development of disease and appropriate therapeutic intervention. To this end, the ability to measure biomolecules in test samples and compare these with given 'norms', taken from healthy individuals (single cell to organism) in a population, is of paramount importance to the management of health and disease. When considering biomolecules it is thus important to include structural variants or anomalies that may arise either spontaneously or as a result of some interaction that can alter functionality. Using advanced bioanalytical tools (such as mass spectrometry, Chapter 9) it is possible to gain both qualitative and quantitative information on a given biomolecule or variant (synthetic or otherwise) which is of scientific and therapeutic importance. This is illustrated briefly below, considering what key classes of biomolecule normally do, what happens when things involving those biomolecules go wrong, and how understanding normal functionality and defects can give new insights into diseases and their treatment.

Nucleic acids (RNA and DNA)

There are various types of ribonucleic acid (RNA) molecule, and some confusion lies in the fact that not all RNA performs the characteristic general function of translating genetic information into proteins. Different RNA molecules have different biological functions: (i) *messenger ribonucleic acid* (mRNA) – carries information from deoxyribonucleic acid (DNA) to ribosomes (cellular protein synthetic 'factories'); (ii) *transfer ribonucleic acid* (tRNA)–transfers specific amino acids to a growing polypeptide chain during protein synthesis (so-called *translation*); (iii) *ribosomal ribonucleic acid* (rRNA) – provides structural scaffolding within the ribosome and catalyses formation of peptide bonds; (iv) *non-coding RNA* (RNA genes) – genes encoding RNA that are not translated into protein; (v) *catalytic RNA*–which catalyses chemical reactions; (vi) *double-stranded ribonucleic acid* (dsRNA) – forms genetic material of some viruses; can initiate *ribonucleic acid interference* (RNAi) and is an intermediate step in *small interfering ribonucleic acid* (siRNA) formation; can induce gene expression at transcriptional level, where dsRNAs are referred to as *small activating RNA*. Problems with the functions of these different RNAs will obviously impact on processes critical to protein synthesis and while, at present, there is little that can be done to fix RNA-driven processes, the ability to detect such alterations is of diagnostic/therapeutic value. For example, as siRNA can knock down specific genes, it has proven experimentally useful in the study of gene function and their role in complex pathways, and also offers the exciting possibility of therapeutic silencing of specific genes mediating disease pathways.

Within DNA lies the genetic code (blueprint) of all living organisms that contains genetic instructions to make individual cells, tissues and organisms. DNA is organized within *chromosomes*, and a set of chromosomes in a cell makes up the cell's *genome*. Furthermore, DNA can replicate to make an identical copy, an important means of transferring genetic information into new cells. While *genes* may be defined in a number of ways, they are generally considered inheritable DNA sequences that both store and carry genetic information throughout the lifespan of an individual. The coding information of genes depends on the bases comprising the DNA, and the sequence of the four bases (i.e. A, T, G, and C) confers the *genetic code* that specifies the sequence of amino acids making up a particular protein within a cell. A process called *transcription* reads the genetic code, where the enzyme RNA polymerase allows transfer of genetic information from DNA into mRNA before the message is translated into protein (translation and protein synthesis). Given the importance of DNA, it is perhaps not surprising that cells inherently have a restricted ability to repair and protect DNA. However, the failure to correct DNA lesions can cause disease, and if *mutated DNA* is *heritable* then it may pass down to offspring. In humans, inherited mutations affecting DNA repair genes have been associated with cancer risk, for example the famous BRCA1 and BRCA2 (which stands for breast cancer 1 and 2, respectively) mutations. Notably, cancer therapy also primarily acts to overwhelm the capacity of cells to repair DNA damage, resulting in preferential death of the most rapidly growing cells, which include the target cancer cells.

Peptides and proteins

Peptides and proteins are often grouped into distinct families according to various criteria, such as structure and primary function. Given that peptides and proteins are major regulators of very many different biological processes, there is an incredibly wide and diverse range of peptides and proteins in nature. Indeed, some peptides/proteins not found in man may still have biological or medical applications in the regulation of human processes (e.g. cell signalling) and related therapeutic applications. For example, a peptide called *exendin* was originally isolated from the saliva of the large, slow moving, venomous lizard, the 'Gila monster' (*Heloderma suspectum*), found in Arizona and other parts of the United States/Mexico. The venom, secreted into the lizard's saliva, contained a rich 'cocktail' of different biological active molecules, including exendin, which was subsequently found to demonstrate antidiabetic properties. While some scientists were initially sceptical about commercial success of this peptide as a pharmaceutical product (under names exenatide or Byetta), it has proven a winner, with >$500 million in sales in its first year! This therapeutic is the first in a new class of medicines which is used to control blood glucose levels in human Type 2 diabetes, and indeed other peptides derived from the human gut peptides glucagon-like peptide-1 (GLP-1) and glucose-dependent insulinotropic polypeptide (GIP) also hold great promise for the future treatment of the 'diabetes epidemic'.

When considering how changes to protein structure can alter function and thus contribute to disease processes, focus should be directed to protein misfolding. This can occur for different reasons, but largely arises due to problems resulting from genetic mutations that can cause defective protein folding, incorrect assembly and processing. Indeed, incorrect folding is associated with defective cellular transport and/or loss of functional activity, which is the molecular basis of a number of diseases. For example, changes in secondary and tertiary protein structure can lead to neurodegenerative disorders. Alterations of so-called prion proteins are closely associated with Creutzfeld–Jakob disease (CJD) (and variant CJD) and transmissible spongiform encephalopathy (TSE). While these diseases have different origins they are related to each other and amyloidoses, as they involve an aberrant accumulation/deposition of proteins as amyloid fibrils or plaques. There are many other examples of diseases arising from protein folding defects in humans, including cystic fibrosis, cataracts, Tay–Sachs disease, Huntington's chorea and familial hypercholesterolaemia, but of course such defects can affect many different species.

Carbohydrates

As noted earlier, there are many different 'types' of carbohydrate which may be grouped according to the number of structural sugar units, or indeed nutritionally. Typically carbohydrates are classified on the basis of the chemical nature of their carbonyl groups and the number of constituent carbon atoms. Carbohydrates represent major fuel sources for most species. However, in addition to being utilized for storage and transport of energy (e.g. starch, glycogen) they also make up structural components in plants (e.g. cellulose) and animals (e.g. chitin). Given such important roles, it is perhaps not surprising that there are a range of disorders associated with incorrect handling (including storage) and usage of carbohydrates, which include: lactose intolerance, glycogen storage disease, fructose intolerance, galactosaemia, pyruvate carboxylase deficiency (PCD), pyruvate dehydrogenase deficiency (PDHA), and pentosuria. This list is by no means exhaustive but it would be amiss not to mention diabetes mellitus, a metabolic disease that has been described as the 'epidemic of the twenty-first century'. Insulin is an important regulator of whole body metabolism and in particular glucose control, where insulin depletion and/or impaired insulin sensitivity of body tissues has a major direct impact on blood glucose levels (glycaemia), often resulting in either hypoglycaemia (glucose too low) or hyperglycaemia (glucose too high). Both states are detrimental if left untreated, and both major forms of diabetes (i.e. Type 1

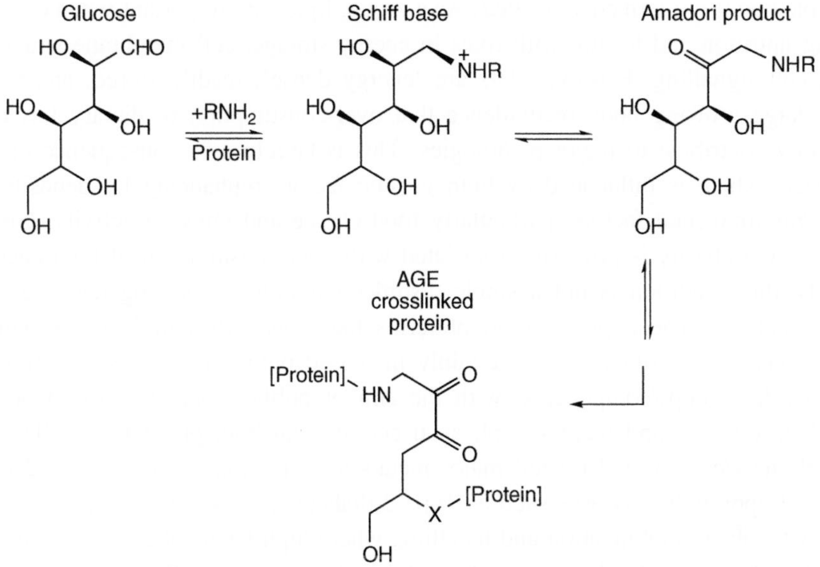

Figure 1.11   **Illustration of advanced glycation end-product (AGE) formation.**

and Type 2) are characterized by hyperglycaemia using various measured parameters including glycated haemoglobin (HbA_{1c}). Glycation is a natural process by which endogenous simple sugars (glucose, fructose, galactose) attach to other biomolecules (typically peptides/proteins) in the bloodstream and in tissues, and subsequently alter their biological activity and often also their elimination from the body. Formation of advanced glycation end-products (AGEs) has three principle steps, the first two of which are reversible; namely, Step 1: Schiff base formation; Step 2: Amadori product formation; and Step 3: Formation of AGE (see Figure 1.11). AGE formation can contribute to major pathologies, especially those associated with Type 2 diabetes and microvascular complications (such as retinopathy, where retinal vascular components are altered/damaged). Epidemiological studies in humans have revealed that tighter control of blood glucose levels (glycaemia) reduces the risk of diabetic complications including those associated with glycation, and that such complications are not an inevitable outcome of diseases such as Type 2 diabetes.

Lipids

Lipids are primarily categorized on the basis of the presence or absence of carbon–carbon double bonds in their carbon chain. When lipids contain double bonds they are described as *unsaturated*, and conversely those that have no double bonds are termed *saturated*. As a group, lipids are important when considering nutrition and health, with roles in energy storage, cell membrane structure and cell signalling. However, fats are 'energy dense', readily stored, and there is a large growing body of evidence that over-consumption of dietary fats can directly contribute to major pathologies. This is largely as a consequence of fat storage, which is influenced by both genetic factors (enhancing fat deposition) and environmental factors, particularly food choice and physical activity. Given this, while obesity is primarily associated with over-consumption of high-energy foods, this condition is not a single disorder but rather a heterogeneous group of conditions. The popularization of terms like *good fats* and *bad fats*, while questionable scientifically, has certainly increased public awareness of different dietary fat groups/components, with the aim of public health benefits. Control of dietary habits and increased physical activity can both prevent and alleviate weightiness/obesity and related major metabolic conditions such as Type 2 diabetes. Importantly, diseases such as Type 2 diabetes are as much about impaired fat metabolism, mobilization and handling, where high levels of circulating lipids combined with high glucose levels collectively result in cell and tissue damage (so-called glucolipotoxicity). Much attention has been directed to cholesterol (and hypercholesterolaemia) and trans fatty acids as important risk factors for heart disease, which itself is associated with obesity/being overweight. Given the emerging obesity epidemic (associated with Type 2 diabetes–diabesity), there is particular pressure on the research community and pharmaceutical industry to discover and develop new drugs to combat obesity. Recent efforts have been directed towards agents that enhance metabolism and in particular fat utilization, or alternatively target the suppression of appetite or regulation of feeding behaviour. These efforts are critically important to avoid the necessity for invasive surgical procedures such as gastric bypass to reverse or prevent worsening of obesity.

Key Points

- Biomolecules are naturally occurring chemical compounds found in living organisms that are constructed from various combinations of key chemical elements.

- Biomolecules can be broadly classified into three main categories: small molecules, monomers and polymers.

- Differences in the properties of biomolecules are dictated by their components, design and construction, giving the inherent key features and characteristics of each biomolecule which enables its specific function.

- Nucleotides consist of three key components: a heterocyclic nitrogenous base, a sugar and one or more phosphate groups.

- Nucleic acids are polymers constructed from nucleotides (monomers), where ribonucleic acid (RNA) comprises ribonucleotides, and deoxyribonucleic acid (DNA) contains deoxyribonucleotides.

- Amino acids are biomolecules that contain a central carbon atom (alpha-carbon) attached to a carboxyl group (COOH), an amine group (NH_2), hydrogen atom (H) and a side chain (R group).

- A peptide is a short molecule formed by amino acids linked through amide (peptide) bonds, while proteins comprise one or more polypeptides.

- Carbohydrates are simple neutral biomolecules composed of C, H, and O, and are classified on the basis of their number of structural sugar units and functional group (aldehyde or ketone).

- Lipids represent diverse classes of hydrocarbon-containing biomolecules, containing polar heads, with important roles in energy storage, cell membrane structure and cell signalling.

- The structure of a biomolecule will confer certain functional attributes which are a key defining feature of that biomolecule.

2 Analysis and quantification of biomolecules

The last chapter considered the main features and characteristics of important biomolecules. This chapter focuses on the core methods used to detect and measure these biomolecules in nature. Quantification of biomolecules lies at the heart of analysis of biological test samples. These samples are key to forensic investigation, clinical tests and research, and come from sources as diverse as soil to body fluids, hair and synthetic fibres. In order to quantify biomolecules within these samples it is necessary to apply a range of technologies, which vary from simple test procedures to analysis with complex state-of-the-art scientific instrumentation. As described, sensitivity, accuracy, and precision are vital in the determination and understanding of the role of individual biomolecules in nature.

Learning Objectives

- To appreciate the importance of accurate determination of biomolecules.

- To outline the principles underlying major methods used to detect and quantify biomolecules.

- To comprehend and apply knowledge of key parameters in the quantification of biomolecules.

- To explain the principles of moles and molarity, and use related equations in basic calculations.

- To distinguish between solubility and dilution, and their application in the preparation of solutions.

Understanding Bioanalytical Chemistry: Principles and applications   Victor A. Gault and Neville H. McClenaghan
© 2009 John Wiley & Sons, Ltd

2.1   Importance of accurate determination of biomolecules

Without the ability to accurately determine the presence and amount of any given biomolecule in nature, we would not be able to understand the contribution and function of these important chemical entities. Since the first basic tests available to distinguish chemical elements, numerous methods have been developed to measure the amounts and types of large and small biomolecules. Individual biomolecules are principally distinguished on the basis of their chemical and physical properties, for example *molecular mass*. In fact, molecular mass is analogous to a 'molecular fingerprint', enabling scientists to specifically distinguish one biomolecule from another using, for example, mass spectrometry techniques that will be considered later. The following basic examples give an immediate illustration of how critical it is to be able to accurately determine the presence and amount of a particular biomolecule in a given sample.

Example 1: Clinical setting

Patient X was hospitalized with chronic fatigue, excessive urination, thirst and progressive weight loss. On first analysis it would seem that a simple urine test could give some indication of the underlying cause. However, while there were elevated levels of glucose in the urine sample, it was impossible to use this test to diagnose and prescribe treatment. This test result prompted an additional measurement of blood glucose levels with the suspicion that the person had diabetes. Only an accurate determination of blood glucose would allow the clinician to confirm whether or not this was the case. A blood sample was taken, and when analysed the glucose level was 16.5 mM; significantly elevated over the normal level, which is <6.0 mM. Without the ability to accurately determine blood glucose it would have been impossible to establish and define the normal levels and subsequently those necessary to define diabetes. Following this test result, blood insulin levels were measured in Patient X, and were extremely low, and thus insulin therapy commenced. If the clinician was unable to accurately determine both blood glucose and insulin levels, Patient X would have remained untreated and may well have entered a dangerous coma as a result of the very high blood glucose levels. Ultimately, if left undiagnosed and untreated this patient would almost certainly have died.

Example 2: Environmental setting

In 1968, control of mosquitoes in Region X necessitated the use of insecticidal chemical DDT (1,1,1-trichloro-2,2-bis(p-chlorophenyl)ethane). At that time DDT and other chlorinated hydrocarbon insecticides had gained popularity due to their effectiveness against a wide range of pests and relatively low mammalian toxicity. However, in 1972, the US Environmental Protection Agency (EPA) banned the use of DDT because of its residual activity and the adverse effects of its accumulation in biological systems. The levels of DDT in Region X were measured in 1997 and were within acceptable limits. However, over the last five years ornithologists had observed a significant decline in the numbers of young birds within the predominant avian species in the area. Further investigations in Region X in 2005 revealed a decrease in eggshell thickness and some of the other hallmark signs of DDT toxicity. Notably there had been a significant increase in the mosquito population several years before. Environmental samples were taken and assessed for DDT and other chlorinated hydrocarbons. These tests revealed significantly elevated levels of DDT compared with the samples taken from the same sites in 1997. Raids of local farms found partially used stocks of DDT insecticide, and the US authorities were able to take action. If it were not possible to have accurately determined that DDT was both present in the environmental samples and subsequent samples of dead birds, it would not have been possible to take appropriate action to stop further dangerous contamination of Region X. As DDT has a reported half-life up to 15 years, further monitoring of Region X, particularly crop species, is necessary to ensure that DDT does not enter the food chain.

There is often some confusion regarding the use of the terms accuracy and precision with respect to the quantification of biomolecules. While **accuracy** is defined by how closely a measured value matches the real or true value, **precision** describes how reproducible the measurements (repeated measurements) are with respect to each other, in other words how closely they match each other (see Figure 2.1).

The level of accuracy and precision required dictate whether or not *qualitative* or *quantitative* analysis is sufficient for the measurement of a particular biomolecule.

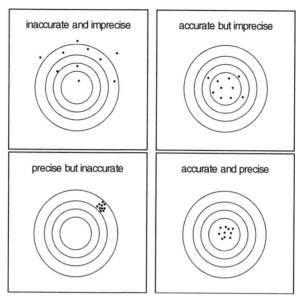

Figure 2.1   Illustration of accuracy and precision using target diagrams.

Qualitative versus quantitative analyses

Qualitative analysis simply determines if a biomolecule is present or not.

Example: A blood sample is taken from an athlete to quickly determine whether or not they have been using any illicit substances.

Quantitative analysis determines the quantity (actual amount or concentration) of a particular biomolecule.

Example: A young person died under suspicious circumstances with suspected alcohol poisoning and there was uncertainty over the validity of the blood specimen, prompting quantitative measurement of alcohol in urine.

Often simple qualitative analysis is followed up by quantitative analyses. A good example of this is allergy testing. When an allergic reaction is suspected it is obviously important to firstly try to identify the presence of the allergen that causes it, before going on to measure the levels of the allergen. This is a logical approach as qualitative analyses are typically cheaper and less time-consuming to perform, and there are obviously too many allergens to justify quantification in the first instance.

Determination of allergen by qualitative followed by quantitative analysis

Qualitative 'skin-prick' test: For this test a drop of fluid containing the suspected allergen is placed on the forearm and a small needle pricks the skin through the drop. If the person tests positive for the allergen there will be signs of redness and itching of the skin around the needle prick and development of a white swelling (or weal). In general, the larger the size of the weal the more likely the person is to be allergic, and diagnosis is confirmed on the basis of clinical symptoms.

Quantitative 'blood' test: For some allergens a skin-prick test is not sufficient. In these cases blood tests are used to determine the level of specific antibodies that target the allergen. The most commonly used technique is the Radio Allergo Sorbent Test (RAST) that measures the amount of immunoglobulin E (IgE) antibodies, which trigger histamine release and the body's immune response.

2.2 Major methods to detect and quantify biomolecules

Numerous *routine* or *specialized* methods are used to detect biomolecules giving either qualitative or quantitative data. The choice of method depends largely on the purpose of the investigative test and the level of accuracy and precision required. However, it is important also to consider other key factors, including the simplicity and time taken to conduct the test (e.g. urgency of getting result), number of tests performed (e.g. hundreds or thousands of samples), and also related direct and indirect costs (e.g. price of test kit, salary of personnel). Cost-effectiveness together with clinical effectiveness may dictate the pros-and-cons of population screening tests for particular biomolecules that are indicators or predictors of a given disease. For example, newborn babies undergo the 'heel-prick' test that measures levels of a number of biomolecules which are markers for disease, including phenylalanine, which would be significantly raised in phenylketonuria (PKU).

Each test method relies on one or more chemical and/or physical attributes of a biomolecule for detection and/or quantification. It is obviously impossible to capture the sheer number of diverse methods that are used in biological, biomedical and environmental testing in this textbook. However, many of these methods

Table 2.1 Basic overview of key principles underlying major bioanalytical methods

Method/instrument	Physical and chemical attributes
Ion-selective electrode	pH and/or ionic charge
Centrifugation	Shape, volume and/or density
Spectroscopy	Energy, molecular structure
NMR	Charge and magnetic properties (e.g. spin/resonance)
Mass spectrometry	Mass
Chromatography	Mass, size (volume/density), shape, charge
Electrophoresis	Mass, charge, shape

utilize specialized instrumentation, which is based on simple chemical or physical principles. The following chapters consider arguably the most important of these techniques to the life and health scientist, but Table 2.1 gives a basic overview of the key principles underlying major bioanalytical methods.

2.3 Understanding mass, weight, volume and density

There are various basic measures that are taken in order to quantify biomolecules. Of these arguably the most important are mass, weight, volume and concentration. Often the term *mass* is confused with that of *weight*, but it is important to recognize that they are actually quite different.

Mass and weight

Mass is the quantity of matter in a given object and is independent of gravity.

Weight is the force of gravity on a given object and is calculated as mass multiplied by gravity.

$$\text{Weight} = \text{mass} \times \text{acceleration due to gravity}$$

While an object will always have a given mass, its weight will vary depending on gravity. So an object with the same mass will have a different weight on Earth compared with the Moon or a weightless environment (such as in space).

Scientific measurements are based on mass (not weight), and the *Systéme International* (SI) unit is the gram (g). For further details on SI units and common prefixes please see Appendix 1.

Determination of mass using common laboratory balances

Somewhat confusingly the means of determining mass is by *weighing*, and the indispensable and versatile device for this is the *analytical balance*. 'Weighings' using such balances are usually made in a 'weighing bottle' or 'weighing boat', and give measures of the mass in air (buoyancy) as opposed to the ideal conditions of measuring mass in a vacuum. Balances make measurements on the basis of comparison of an unknown against a known weight (or mass), and the most common analytical balances are electronic or mechanical single-pan. Although the latter are still excellent balances they have largely been superseded by electronic balances, which are more user friendly and suffer from fewer errors and mechanical failures (see Figure 2.2).

For the most part, bioanalytical chemistry considers small masses in the order of grams to micrograms, and standard laboratory weighings are typically made to three or four significant figures (e.g. 0.0002 g). Obviously accuracy and sensitivity are important for small measures, and analytical balances have different weighing *ranges* and *readabilities*. For example, a macrobalance may have a range of 160 g

Figure 2.2   **Photograph illustrating typical laboratory balances and weighing equipment.**

and readability of 0.0001 g, while a semi-microbalance may have a range of 30 g and readability of 0.00001 g. Following the same pattern, a microbalance may have a range of 1 μg, and an ultramicrobalance 0.1 μg.

Rules and errors in determination of mass

There are various sources of error that can affect the accurate measurement of mass of a given substance when using an analytical balance. These errors largely relate to changes in environmental conditions, which may fluctuate in a given setting or differ from one laboratory to another. To minimize such errors, levelling tables, anti-vibrating balance benches or mats, and static controllers can be used, and other factors such as temperature and humidity are often also controlled. Care should also be taken in order to maintain good working order of analytical balances, avoiding contamination, dust, and corrosion. There are also basic common-sense rules when weighing, including the use of clean spatulas, and weighing bottles/boats, avoiding spillage onto the pan; it is also important to avoid air currents, which can be minimized by closing the door of the analytical balance and allowing sufficient time for the measurement to be made.

Volume and density

Volume is the space that mass in a given object occupies.

Density is a comparison of an object's mass to its volume.

$$\text{Density} = \text{mass/volume}$$

It is no more correct to assume that a small object will have a low mass, than it is to assume that an object with a large volume has a high mass. Density, however, gives an indication of 'how big' something actually is, as each substance has a unique density. The above formula dictates that objects of high density have a lot of mass occupying a relatively small volume.

Example: The densities of gold and lead are $0.0193\,\text{g}\,\text{l}^{-1}$ and $0.0113\,\text{g}\,\text{l}^{-1}$, so if you have a 200 g bar of each, which will occupy the most volume?

Substituting these values into the above formula:

Gold: $0.0193 = 200/\text{volume}$

Therefore, cross-multiplying, $\text{volume} \times 0.0193 = 200$

So, volume = 200/0.0193 = 10 362.691

Lead: 0.0113 = 200/volume

Therefore, cross-multiplying, volume × 0.0113 = 200
So, volume = 200/0.0113 = 17 699.111
From the above you will see that lead occupies a considerably greater volume than gold.

As shown in the above example, the SI unit for volume is litres (l) and density is grams per litre ($g\,l^{-1}$). For further details on SI units and common prefixes, please see Appendix 1.

Basics of volume measurement

Before measuring a volume it is important to choose the most appropriate equipment in order to achieve the greatest accuracy. The volume and level of accuracy will help determine which piece of equipment should be used. The most common equipment includes various pieces of glassware (e.g. volumetric flask, measuring cylinder, burette, pipette), mechanical micropipette (or pipettor) and syringes (Figure 2.3).

While both a measuring cylinder and volumetric flask are useful in the measurement and dispensing of larger volumes (typically 5–5000 ml), a volumetric flask has a higher level of accuracy. Conical flasks or beakers are particularly convenient for dispensing liquids, but as they have a much lower accuracy they should not be used for measurement of volume.

As with the use of an analytical balance, there are situations where accuracy is required when dispensing liquids and measuring volumes. The importance of accurate volumetric measurement cannot be overemphasized, particularly when preparing solutions following quantitative calculations. Given this, it is relatively pointless accurately weighing a solid using an analytical balance if the same degree of care is not used when measuring the volume into which it is dissolved. Even a small change in volume can have a serious impact on the final concentration and other variables such as pH. This error can become particularly significant if small volumes of highly concentrated solutions are required. For example, if 2 g of X are dissolved in 10 ml this gives a $0.2\,g\,ml^{-1}$ concentration, however, if the same 2 g of X are dissolved in 11 ml this gives a $0.18\,g\,ml^{-1}$ concentration—a difference of 10%!

Figure 2.3   **Typical pieces of laboratory glassware used in volume measurement.**

Use of a micropipette

Micropipettes, also referred to as *autopipettors* (or *pipettors*), are key tools for the bioanalytical chemist. These are used to very accurately and precisely measure small volumes ranging from 1 to 1000 μl. While there are various manufacturers, most micropipettes have the same standard features and are either single channel or multichannel, and may be semi- or fully automatic (see Figure 2.4).

2.4   Understanding moles and molarity

In order to understand the principles of *moles* and *molarity* it is important to have a basic appreciation of the terms associated with atoms, atomic structure and molecules. These terms are often perceived as confusing or complicated, and the following subsections are aimed at demystifying moles and molarity.

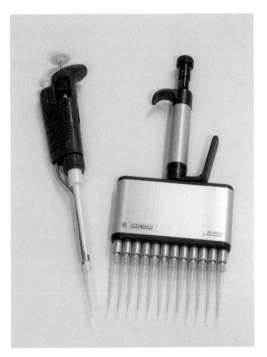

Figure 2.4   **Photograph showing a single channel and multichannel Gilson Pipetman.**

The Mole
The **mole (mol)** is the SI unit of a quantity of a given molecule.

Origin of the term mole

The term *mole* is the English version of the German word for molecular weight ('mol', short for *molekulargewicht*, rather than referring to the burrowing mammal). Chemists who coined this term around 200 years ago were trying to quantify the chemical composition of gases, solids and liquids. For this they needed a universal unit of measurement so they could connect the mass of an atom (or molecule) with the number of atoms or molecules in any given compound. They decided to use Avogadro's number.

What is a mole?

A mole is a term that represents a number; just like the word 'dozen' represents '12', the word 'mole' represents 6.02×10^{23} (or 602 216 900 000 000 000

000 000). The term *mole*, while initially confusing, is extremely useful. To give an example: if we were to count the number of water molecules in a single drop it would be around 10 trillion (10 000 000 000 000), so rightly or wrongly chemists argue that it is easier to quantify such numbers in moles and thus cut down on the number of zeros!

Why 6.02 × 10^{23}?

This number is 'Avogadro's number', named after the Italian physics professor who proposed that equal volumes of different gases at the same temperature contain the same number of molecules. In its simplest form, 1 mol of any substance contains 6.02×10^{23} atoms or molecules of that substance, and as this relationship was first put forward by Avogadro they named this number after him.

Concept of the mole

A defined mass of an element (its atomic weight) contains an exact number of atoms, that is, Avogadro's number. Therefore, for any given molecule, one mole of the substance has a mass (in grams) equal (numerically) to the atomic mass of the molecule. One mole is the number of atoms in exactly 12 thousandths of a kilogram (i.e. 12 g) of carbon (^{12}C), which is the most abundant isotope of carbon.

Example: A single molecule of silver has an atomic mass of 107.8682; therefore 1 mol of silver weighs 107.8682 g.

Now we have considered the basic definition of the mole it is important to gain insights into the use of moles in fundamental formulae and equations.

The following is the most basic formula used to calculate the number of moles in a given substance:

$$\text{Number of moles} = \frac{\text{mass of substance}}{\text{relative formula mass (RFM)}}$$

The *relative formula mass* (RFM) equals the atomic or molecular weight of the substance, so for a compound such as sodium chloride (NaCl) this is simply the sum of the atomic weights of Na (23) and Cl (35.5), taken from the Periodic Table of the elements (see Appendix 2). The terms 'relative formula mass' and 'relative atomic mass' are often taken as interchangeable.

Worked examples

Example 1: Calculate the number of **moles** contained in 5 g of sodium chloride, NaCl, which has a RFM of 58.5.
Using the equation:

$$\text{Number of moles} = \frac{5}{58.5} = 0.085$$

Example 2: Calculate the number of **moles** contained in 50 g of calcium hydroxide, $Ca(OH)_2 \cdot 6H_2O$, which has an RFM of 182.
Simplified, $Ca(OH)_2 \cdot 6H_2O$ contains 1 Ca, 8 O and 14 H, so the RFM is calculated as the sum of the individual parts: 40 (1 × Ca) plus 128 (8 × O) plus 14 (14 × H), which equals 182.

$$\text{Number of moles} = \frac{50}{182} = 0.275$$

Using the number of moles and the RFM we can also use this formula to calculate mass.

Example 3: Calculate the mass of 10 **moles** of calcium chloride, $CaCl_2$, which has a RFM of 111.
Simplified, $CaCl_2$ contains 1 Ca, 2 Cl, so the RFM is the sum of the individual parts: 40 (1 × Ca) plus 71 (2 × Cl), which equals 111.

$$10 = \frac{\text{mass}}{111}, \text{ therefore mass} = 10 \times 111 = 1110, \text{ or } 1.11 \times 10^3 \text{ g}$$

When doing calculations it is important to watch the units. In this formula the mass must be given in grams (g), therefore if the mass is recorded as milligrams (mg) it is necessary to convert this to grams for the calculation.

Example 4: Calculate the number of **moles** in 250 mg of sodium chloride, NaCl, which has a RFM of 58.5.
The mass of NaCl in grams = 0.250
Using the equation:

$$\text{Number of moles} = \frac{0.250}{58.5} = 0.00427 = 4.27 \times 10^{-3}$$

Example 5: Calculate the number of milligrams in 3×10^{-5} **moles** of calcium chloride, $CaCl_2$, which has a RFM of 111.

The number of moles $= 0.00003$

$$0.00003 = \frac{\text{mass}}{111}, \text{ therefore mass} = 0.00003 \times 111 = 0.00333\,\text{g}$$

Mass in grams is 0.00333 (or 3.33×10^{-3}), which is $3.33\,\text{mg}$.

Just like appreciating that a mole of a compound is the number of grams equal to its molecular mass, molarity is another important measure that is fundamental to bioanalytical chemistry.

Molarity

The **molarity** of a solution is the number of moles of dissolved substance (solute) in a given volume of liquid (solution).

Importance of molarity

Molarity is a fundamental way of expressing the concentration of a compound in solution. A one molar solution (denoted $1\,\text{M}$) contains one mole of a given substance in a litre of a solution. In other words, $1\,\text{M}$ is equivalent to 1 mole per litre ($1\,\text{mol}\,\text{l}^{-1}$), and is a convenient means of representing the concentration of substances in all liquids and body fluids including blood plasma, urine and cerebrospinal fluid.

Now we have considered the basic definition of molarity, it is important to gain insights into the use of molarity in fundamental formulae and equations.

In its most basic form molarity is given as follows:

$$\text{Number of moles} = \text{molarity} \times \text{volume (in litres)}$$

Worked example

Calculate the **molarity** of $0.05\,\text{mol}$ of a $250\,\text{ml}$ solution of sulfuric acid, H_2SO_4.

Using the equation:

$$0.05 = \text{molarity} \times 0.250$$

$$\text{molarity} = \frac{0.05}{0.250} = 0.2\,\text{M}$$

As bioanalytical samples are usually measured in millilitres it is more convenient to use the following equation:

$$\text{Number of moles} = \frac{\text{molarity} \times \text{volume (in millilitres)}}{1000}$$

Therefore:

$$\text{Number of moles} = \frac{\text{mass of substance}}{\text{RFM}} = \frac{\text{molarity} \times \text{volume (in millilitres)}}{1000}$$

Rearranging this, we get arguably the most important formula of all for bioanalytical chemists, used routinely in clinical and research laboratories. It is important for all life and health scientists to be able to remember and use this formula:

$$\text{Mass of substance} = \frac{\text{molarity} \times \text{volume (in millilitres)}}{1000} \times \text{RFM}$$

Worked examples

Example 1: Calculate the number of grams of sodium sulfate, Na_2SO_4 (RFM 142), which should be weighed out to prepare 250 ml of a 0.15 M solution. Using the equation:

$$\text{Mass of substance} = \frac{\text{molarity} \times \text{volume (in millilitres)}}{1000} \times \text{RFM}$$

$$\text{Mass of substance} = \frac{0.15 \times 250}{1000} \times 142 = 0.0375 \times 142 = 5.325\,\text{g}$$

Example 2: Calculate the molarity of 350 ml of a solution containing 4.5 g of sodium sulfate, Na_2SO_4 (RFM 142).

$$\text{Mass of substance} = \frac{\text{molarity} \times \text{volume (in ml)}}{1000} \times \text{RFM}$$

$$4.5 = \frac{\text{molarity} \times 350}{1000} \times 142$$

$$\frac{4.5 \times 1000}{350} = \text{molarity} \times 142$$

$$12.857 = \text{molarity} \times 142$$

$$\text{Molarity} = \frac{12.857}{142} = 0.091\,\text{M}$$

Example 3: Calculate the volume required to make a 0.5 M solution containing 6.2 g of sodium sulfate, Na_2SO_4 (RFM 142).

$$\text{Mass of substance} = \frac{\text{molarity} \times \text{volume (in millilitres)}}{1000} \times \text{RFM}$$

$$6.2 = \frac{0.5 \times \text{volume (in millilitres)}}{1000} \times 142$$

$$\frac{6.2 \times 1000}{0.5} = \text{volume} \times 142$$

$$12\,400 = \text{volume} \times 142$$

$$\text{Volume} = \frac{12\,400}{142} = 87.32\,\text{ml}$$

Practical considerations when preparing solutions

As outlined above, solutions are prepared on the basis of their molar concentrations. In order to prepare a solution of known concentration from a solid material it is important to know and decide upon the final concentration required, together with the final volume of solution. Also, it is important to note that a solution such as a buffer may contain more than one dissolved substance and using the above equation it is possible to use separate calculations to determine the mass of all given solids needed for a particular solution.

Example: You are required to make a 150 ml solution containing 0.4 M NaCl (RFM 58.5) and 0.6 M $CaCl_2$ (RFM 111). What mass of each salt must be weighed out for the preparation of this solution?
There are two separate calculations here: (i) mass of NaCl required; (ii) mass of $CaCl_2$ required.

$$\text{Mass of NaCl} = \frac{\text{molarity} \times \text{volume (in millilitres)}}{1000} \times \text{RFM}$$

$$= \frac{0.4 \times 150}{1000} \times 58.5 = 3.51\,g$$

$$\text{Mass of CaCl}_2 = \frac{\text{molarity} \times \text{volume (in millilitres)}}{1000} \times \text{RFM}$$

$$= \frac{0.6 \times 150}{1000} \times 111 = 9.99\,g$$

In order to make this solution, both salts should be weighed accurately using a weighing boat and transferred to a beaker. A volume less than 150 ml (let's say 100 ml) is added to the beaker and the salts allowed to fully dissolve, stirring and heating as required. This solution is then fully transferred into a volumetric flask where the volume is then accurately made up to the 150 ml mark.

Other terms used when considering solutions

Osmolarity: Where *molarity* is the term used to express the concentration of a solution based on the number of moles per litre, *osmolarity* is the term used to express the concentration of a solution based on the number of osmotically active moles per litre. The unit of osmolarity is expressed as osmoles per litre.

Example: A 0.25 M solution of NaCl has an osmolarity of 0.5 osmol l^{-1}; this comes from the fact that there are two osmotically active particles (Na^+ and Cl^-) in this solution.

Molality: Where *molarity* is the term used to express the concentration of a solution based on the number of moles per litre, *molality* is the term used to express the concentration of a solution based on number of moles of a solute (dissolved substance) per mass of solvent. The unit of molality is moles per kilogram (or molal).

Example: What is the molality of a solution containing 20.0 g of bromine (Br_2; RFM 159.8) dissolved in 1.5 l of cyclohexane (density 0.779 kg l^{-1})?
   Number of moles of Br_2 is 0.125 (i.e. 20 divided by 159.8).
   To determine the mass of cyclohexane based on density (i.e. mass = density $\times$ volume) we multiply 0.779 $\times$ 1.5, which equals 1.169 kg.
   Therefore the molality is calculated as moles of Br_2 divided by mass of cyclohexane (i.e. 0.125/1.169), which equals 0.107 mol kg^{-1} (or 0.107 molal).

Osmolality: Where *molality* is the term used to express the concentration based on number of moles of a solute per mass of solvent, *osmolality* is the term used to express the number of moles of osmotically active solute particles per mass of solvent. The unit of osmolality is osmoles per kilograms.

Example: What is the osmolality of a $2.0 \, mol \, kg^{-1}$ solution of NaCl which has an osmotic coefficient (ø) of 0.983 (at 25 °C)?

The osmotic coefficient is a set value which is used as a correction factor, and as it varies depending on the molality it must be taken from set values given in tables.

Osmolality is calculated as molality ($2.0 \, mol \, kg^{-1}$) multiplied by the number of solute particles (*n*), in this case 2 (one from Na and one from Cl), multiplied by the osmotic coefficient (ø; 0.983), which equals $3.932 \, osmol \, kg^{-1}$.

2.5   Understanding solubility and dilutions

The terms solubility and dilution are fundamental to laboratory chemistry, but the ability of substances to dissolve (their solubility) and become more dilute (less concentrated) in body fluids is important when considering environmental exposure to certain agents, and in the oral administration of drugs. In both cases, solubility helps to establish the distribution or localization of a particular substance in the body. Also, during this process the particular properties of a substance may enable it to become widely distributed (diluted) throughout the body or instead be concentrated in a particular tissue or organ. As such, the solubility of a drug such as aspirin plays an important part in delivering the active component(s) of this drug to the site of action or target tissue.

Solubility

The **solubility** is the amount of a particular substance (solute) that can dissolve in a given solvent (giving a saturated solution) at a particular temperature. Given this, solubility is critically dependent upon two main factors, namely, type/properties of **solvent** (e.g. organic or inorganic) and **temperature** (e.g. 12 °C or 37 °C, sub-physiological and physiological temperature).

Solvent: Water is the most abundant solvent, and while a substance may be soluble in water it may not be soluble in other solvents (hence the popular terms water-soluble or fat-soluble).

Temperature: In general, increasing the temperature increases the solubility up to the *point of saturation*, which is when no more substance can be dissolved in that particular volume of solvent, and excess solute may be visible in suspension.

Solubility can be expressed in several ways, the most common being molarity (as moles of solute per litre) and % solubility, which is the number of grams of solute dissolved in 100 g (or 100 dm^3) of solvent.

To confuse matters, the solubility of a substance is often defined on its ability to 'go into solution' at 25 °C (i.e. around room temperature). On the basis of this definition, whereas soluble substances can form a 0.10 to 1 M solution at 25 °C, insoluble substances cannot form a 0.10 to 1 M solution at 25 °C.

Importance of solubility

Water solubility (or aqueous solubility) is the maximum amount of a given substance that can dissolve in pure water at a particular temperature. Water solubility is a critical parameter, as drugs hitting the stomach must dissolve (be soluble) in gastric fluids to enable their absorption, transport, and delivery to target organs (e.g. the liver). The solubility of a substance in water is also helpful in determining the dispersion and fate of chemicals in the environment, and for this, water solubility is related to other chemical parameters (e.g. the octanol–water partition coefficient; KOW).

In addition to solubility, the concentration of a dissolved substance in a given solvent can be expressed in 'parts per' notation, when low concentrations of a particular substance are still considered significant. In these cases the metric system is the most convenient way to express values as metric units increase stepwise as ten, hundred, thousand and so on. Typically this is given on the basis of mass but it can also be expressed by volume, mass/volume (m/v) ratio, or number of moles. For example: A milligram (1 mg or 0.001 g) is one 'part per thousand' of a gram (1000 mg or 1 g) or one 'part per million' (ppm) of a kilogram (1 000 000 mg or 1000 g).

Worked example

Zinc (Zn) is an essential element for many plant functions. A 2 g sample of sweet corn was analysed for Zn content, which was measured as 6.8 μg.

Calculate the concentration of zinc (Zn) in both parts per million (ppm) and parts per billion (ppb) in this sample.

$$\frac{\text{Mass of Zn}}{\text{Mass of sweet corn}} = \frac{6.8\,\mu g}{2\,0000\,0000\,\mu g} = 0.000\,0034 = 3.4\,\text{ppm}$$

Since 1 ppm = 1000 ppb

$$3.4\,\text{ppm} = 3.4 \times 1000\,\text{ppb} = 3400\,\text{ppb}$$

Note that ppm is milligrams per kilogram or micrograms per kilogram and ppb is micrograms per kilogram or nanograms per kilogram.

'Parts per' notation is a useful way to express the relative abundance of trace elements in earth, forensic samples, or the environment, and this notation is particularly relevant when considering the levels of pollutants/contaminants in any system.

Example: Mercury is a known contaminant of seafood, and regulatory bodies such as the US Food and Drug Administration (FDA) recommend consumption of certain fish on the basis of their ppm levels of methylmercury substances, with a set 'action level' for mercury in seafood of 1 ppm. The US EPA has also set an official reference dose (RfD) for mercury at $0.1\,\mu g\,kg^{-1}$ per day in humans (corresponding to a blood mercury level of $5.8\,\mu g\,l^{-1}$ or 5.8 ppb), which is defined as the amount of mercury an individual (including sensitive subpopulations) can be exposed to on a daily basis over their lifetime without appreciable risk of effects.

Dilution

Dilution is the process of making a weaker solution, which has a lower concentration of a given solute, and is fundamental when preparing solutions.

So-called **stock solutions** are solutions typically 10, 100 or 1000 times more concentrated than that ultimately required in the final solution. Stock solutions are particularly useful when the same ingredients are required in multiple test solutions, when various concentrations of these ingredients are required or simply for storage purposes. A useful example is fruit-flavoured drink concentrates (syrups or cordials) that are mixed/diluted with water to taste; if these drinks were pre-diluted they would typically fill hundreds of bottles. The same principle applies to the preparation and storage of chemical concentrates for preparation of laboratory reagents, solutions and buffers.

Making a dilution

The concentration is the amount of substance per volume of solution (e.g. millimoles per litre; millimolar) and in order to conduct the dilution it is critical to first determine the concentrations and volumes of the original and final solutions.

For all dilutions the concentration of the original solution (C_1) will always be less than that of the final solution (C_2), and the volume of the original solution (V_1) will be less than that of the final solution (V_2). Through diluting with, for example, water we are increasing the volume by adding to the original solution, while at the same time decreasing its concentration. Thus the following relationship holds:

$$C_1 V_1 = C_2 V_2 \quad \text{sometimes given as} \quad [C_1]V_1 = [C_2]V_2$$

Hint: To avoid simple errors it is best to always express both C_1 and C_2 in molar (i.e. moles per litre) and V_1 and V_2 in litres.

Example 1: How much of a 2 mM solution of NaCl is required to prepare 3 l of a 1 mM solution?

$$0.002 \times V_1 = 0.001 \times 3$$

$$V_1 = \frac{0.001 \times 3}{0.002} = 1.5\,l$$

Example 2: How much of a 20 mM solution of NaCl is required to prepare 0.5 l of a 1 mM solution?

$$0.02 \times V_1 = 0.001 \times 0.5$$

$$V_1 = \frac{0.001 \times 0.5}{0.02} = 0.025\,l = 25\,ml$$

Dilution factors and series

Dilution factor: The ratio between the initial concentration of stock (starting) solution and the concentration of the final solution.

Example: Calculate the dilution factor when preparing a final solution of 100 ml of 0.4 M from a 10 ml stock solution of 8.0 M concentration.

$$\text{Dilution factor} = \frac{\text{final concentration required}}{\text{concentration of stock solution}}$$

$$= \frac{0.4}{8.0} = \frac{1}{20}$$

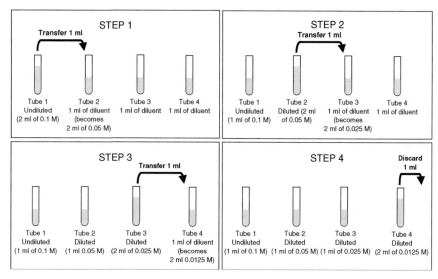

Figure 2.5   **Diagrammatic representation of a stepwise serial dilution.**

This ratio is commonly referred to as a *fold* dilution, in this case a 20-fold dilution.

To make the final solution (i.e. 100 ml of 0.4 M) only a portion of the stock solution is required, in this case 1/20 of 100 ml, which equals 5 ml. In order to prepare the final solution, 5 ml of the stock solution should be accurately transferred to 95 ml of diluent, making the final required solution, which is 100 ml and 0.4 M (where 5 ml in 100 ml equals 1/20).

Dilution series: Comprises a range of solutions with different dilution factors, usually representing a stepwise decrease in concentration (see Figure 2.5).

Key Points

- Accuracy is defined by how closely a measured value matches the real or true value, while precision describes how reproducible the measurements are with respect to each other.

- Qualitative analysis simply determines if a biomolecule is present or not, whereas quantitative analysis determines the quantity (actual amount or concentration) of a particular biomolecule.

- Methods used to detect and quantify biomolecules rely on one or more chemical and/or physical properties of that particular biomolecule.

- Mass is the quantity of matter in a given object and is independent of gravity, whereas weight is the force of gravity on a given object and is calculated as mass multiplied by gravity.

- Volume is the space that mass in a given object occupies, whereas density is a comparison of an object's mass to its volume.

- Mole is referred to as the amount of substance containing Avogadro's number of particles.

- Molarity of a solution is the number of moles of dissolved substance (solute) in a given volume of liquid (solution).

- Solubility is the amount of a particular substance (solute) that can dissolve in a given solvent (giving a saturated solution) at a particular temperature.

- Dilution is the process of making a weaker solution that has a lower concentration of a given solute, and is fundamental when preparing solutions.

- Dilution factor is the ratio between initial concentration of stock (starting) solution and the concentration of the final solution, while a dilution series is a range of solutions with different dilution factors.

3 Transition metals in health and disease

Transition metals comprise part of the d-block (groups 3–11) or middle portion of the Periodic Table of elements (see Appendix 2). The term *transition metal* arose as a result of their position in the Periodic Table and how they represent the transition between group 2 through to group 12 elements. Transition metals are widely distributed throughout the earth and oceans, and play extremely important roles in nature. While transition metals are considered 'trace elements' in mammals, this by no means reflects their importance, and these metals play a key, often essential, role in many biological processes and, in particular, the catalysis of physiological enzymatic reactions. This chapter considers some of the core features of transition metals and their role in the regulation of normal physiological processes and the pathogenesis of disease.

Learning Objectives

- To describe and explain the structure and characteristics of key transition metals.

- To outline and discuss the importance of transition metals in physiological processes.

- To appreciate and convey the role of transition metals in disease processes.

- To give examples to illustrate the therapeutic implications of transition metals.

- To demonstrate knowledge of how transition metals are determined in nature.

Understanding Bioanalytical Chemistry: Principles and applications Victor A. Gault and Neville H. McClenaghan
© 2009 John Wiley & Sons, Ltd

3.1   Structure and characteristics of key transition metals

By the IUPAC (International Union of Pure and Applied Chemistry) definition, a *transition metal* is an element, an atom of which contains an incomplete d shell, or that gives rise to a cation with an incomplete d shell. Transition metals have a total of nine atomic orbitals, but only some of these are used for bonding to *ligands*. *Coordination number* denotes the number of donor atoms associated with the central atom and dictates the shape, or *stereochemistry*, of the *coordination complex*. The most common coordination numbers for the first transition metal series are four (tetrahedral and square-planar) and six (octahedral). Metals in the early second and third series are larger and thus can have higher coordination numbers, forming more structurally complex molecules.

It's all about electrons

The electronic structure of transition metals in their *ground state* and *oxidation states* often causes considerable confusion. As electrons are negatively charged particles, removal of electrons makes the metal more positive (hence + sign), and conversely when electrons are added to a metal it becomes more negative (hence–sign). When considering metals and electrons there are three important terms to remember:

Oxidation: This describes the process by which electrons are removed from the metal (M), where the metal has been oxidized. Notably solutions containing Fe(II) tend to oxidize to the Fe(III) ion.

Reduction: This describes the process by which electrons are added to the metal (M), where the metal has been reduced.

Oxidation state: This refers to the number of electrons that have been added to or subtracted from the metal. It is often given in brackets, and where one electron is removed from M, the oxidation state is +1 (or $^+$), denoted as M(I), whereas addition of two electrons to M places it in the −2 (or 2^-) oxidation state, denoted as M(−II).

The ability of transition metals to adopt a number of different oxidation states is a very important feature of transition metal chemistry. In order to determine the oxidation state of a metal in a complex, a simple formalism is adopted where the oxidation state is 'the charge remaining on the central metal atom when all the ligands are removed in their closed shell configuration'. While this

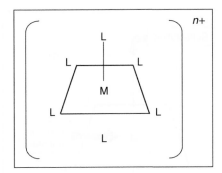

Figure 3.1 **Simple octahedral metal complex.**

formal oxidation state has little to do with the actual charge on the metal or physical properties of a complex, it does help. Another generalized assumption is that transition metal complexes are either *ionic* or *covalent*, as are described by the *crystal field theory* and *molecular orbital theory*, respectively. Crystal field theory provides perhaps the most commonly used description for bonding in transition metal complexes, assuming that it is purely ionic.

Transition metals are integrated into complex structures consisting of a central metal ion or atom surrounded by a set of ions or molecules called *ligands*. Figure 3.1 shows a simple octahedral (eight-sided) metal complex with defined geometry consisting of a transition metal (M) surrounded by a number of ligands (L) and oxidation state (i.e. $n+$). Convention dictates that when a ligand is removed from a metal it takes with it both electrons from the *coordinate bond*, and while some ligands are removed as neutral molecules (e.g. H_2O and NH_3), others are removed as ions (e.g. chloride ions, Cl^- or hydroxide ions, OH^-).

The reactions outlined in Figure 3.2 can occur at different rates, from so-called 'instantaneous' to taking up to several days, and are each considered briefly in turn.

(a) *Substitution:* Replacement by removal of one ligand and addition of another (X with Y). There are two main types of substitution reaction, *associative* and *dissociative*, with different kinetics, which are demonstrated by the simple tetrahedral metal complex shown in Figure 3.3a. However, these apparently simple reactions have further complexity, for example a substitution reaction at a square-planar site can have intermediate steps (solvent-intermediate; Figure 3.3b), and the stereochemistry of the complex is also important (i.e. trans complexes give trans products and cis complexes give cis products, exclusively; Figure 3.3c). These factors also have an influence on substitution reactions of other complexes such as the octahedral metal complex considered earlier (Figure 3.3d).

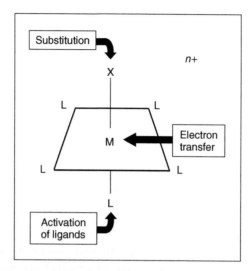

Figure 3.2   Three principal reactions associated with transition metal chemistry, which combine to make up other important processes.

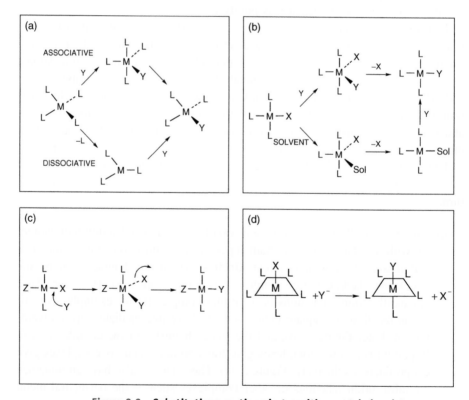

Figure 3.3   Substitution reactions in transition metal chemistry.

(b) *Electron transfer:* Change in the oxidation state of the transition metal (e.g. M(III) to M(II)). These reactions are a major feature of transition metal chemistry and in defining the characteristics of this area of the Periodic Table. The mechanisms by which electrons are transferred between two metal complexes, the so-called *inner-sphere* pathway (where reactants share a ligand during electron transfer) and *outer- sphere* pathway (where reactants do not share a ligand during electron transfer) have been the focus of much study. At its simplest, electron transfer from one transition metal to another is given in Figure 3.4a, but an obvious consideration for electron transfer is the distance the electron has to travel from one species to another. When the reaction is thermodynamically favourable and the two species approach each other closely then electron transfer takes place generating a *redox potential* ($E°$) (Figure 3.4b). Different redox potentials reflect the different oxidation states of the metal.

(c) *Activation of ligands:* Chemical attack of one or more of the ligands (e.g. by hydroxide). In addition to considering the role of the transition metal in a given complex, it is also important to appreciate the general reactivity of ligand(s). For example, knowledge of the effects of reactive species such as hydroxide (OH^-) on organic carbonyl compounds has greatly enhanced understanding of the mechanisms of action of the zinc-containing enzymes, carboxypeptidase and carbonic anhydrase. A number of transition metal ions such as cobalt (Co^{2+}) or copper (Cu^{2+}) play an important role in catalyzing *hydrolysis* of amides, esters and peptides, illustrated in Figure 3.5.

Figure 3.4   Electron transfer in transition metal chemistry.

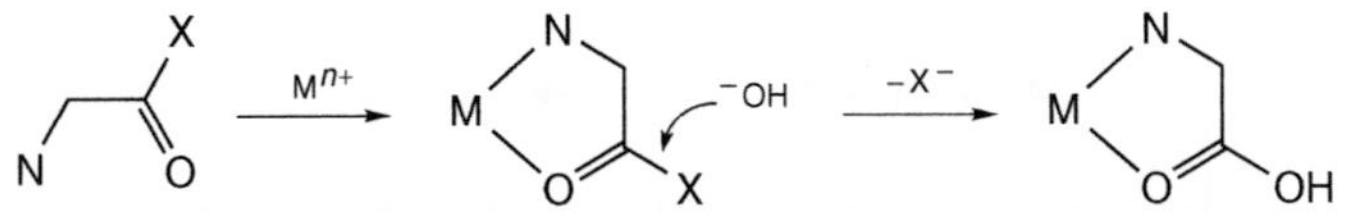

Figure 3.5   Activation of ligands in transition metal chemistry.

In the case of carbonyl compounds, binding of the metal ion to a peptide pulls electron density away from the *carbonyl carbon atom*, making it more susceptible to hydroxide attack.

(d) *Redox-catalyzed substitution:* A combination of (a) and (b). Changing the oxidation state of a transition metal, for example, chromium from III to II, can radically alter the features of the metal reaction site, increasing the rapidity of the reaction. So by making use of this redox property of transition metal complexes, sluggish chromium(III) reactions can be accelerated by addition of catalytic chromium(II) species (Figure 3.6).

(e) *Oxidative-addition reaction:* A combination of (a) and (b). This is where there is a simultaneous increase in oxidation state of the transition metal and its coordination number; simply illustrated in Figure 3.7. In this instance $Y - Z$ could be any of a range of molecular species, the most popular example of which is the alkyl halides (RBr, RCl or RI). The reactivity of alkyl halides to a given metal complex is determined by the nucleophilicity (relative ability of the nucleophile to undergo a reaction) of the transition metal site, and is in the order $RI > RBr > RCl$.

(f) *Insertion:* A combination of (a) and (c). In this type of reaction small molecules such as carbon monoxide, nitric oxide and alkenes are inserted into metal–carbon bonds and sometimes metal–hydrogen bonds. However, despite the title, insertion mechanisms are not straightforward. For example, the attacking molecule may not actually be the group that is undergoing the

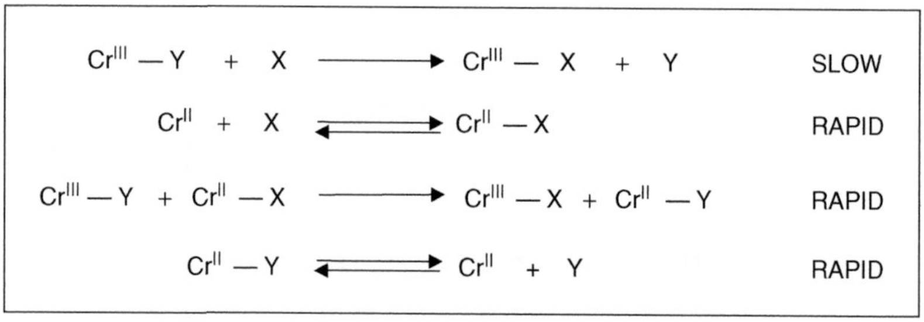

Figure 3.6   Redox-catalyzed substitution in transition metal chemistry.

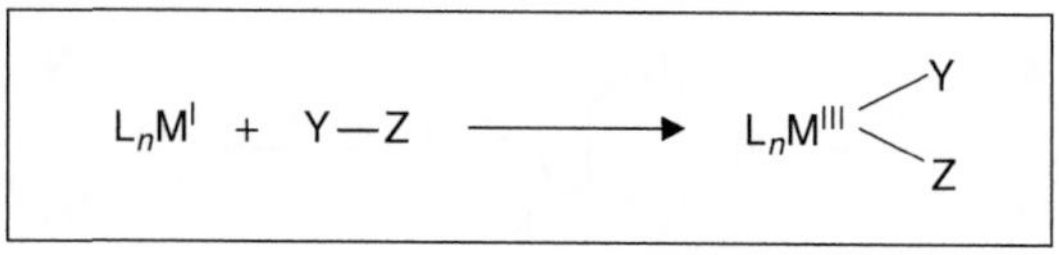

Figure 3.7   Oxidative addition in transition metal chemistry.

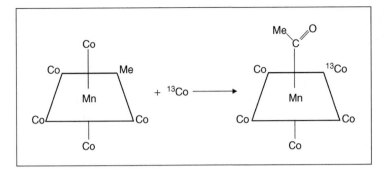

Figure 3.8   Insertion in transition metal chemistry.

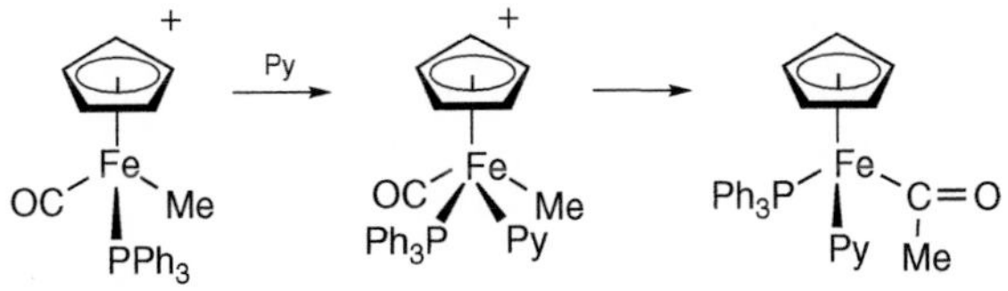

Figure 3.9   Redox-catalyzed insertion in transition metal chemistry.

insertion reaction. This is illustrated in Figure 3.8, where it is one of the molecules already coordinated to the manganese that undergoes the insertion reaction, not the radiolabelled attacking carbon monoxide molecule (^{13}CO).

(g) *Redox-catalyzed insertion:* A combination of (b) and (c). Transient electrochemical techniques have demonstrated that in some instances there is a redox-induced migratory insertion reaction where a molecule (for example pyridine, py) can bind to a metal centre and catalyze migration of another group (e.g. a methyl group). This example is illustrated in Figure 3.9.

Other key characteristics of transition metals

Amplification of trace element action: As the action is amplified, only small (milligram) amounts of transition metals are necessary for optimal physiological performance. For example, lack of even a small amount of iron can result in a clinical abnormality such as anaemia, which seems extreme relative to the deficit in the transition metal.

Specificity: Essential transition metals are specific for their own physiological functions, that is, they cannot be directly replaced by chemically similar elements.

Homeostasis: The mechanisms ensuring optimal absorption, distribution, storage and excretion of an element over a range of intakes constitute an effective system of homeostatic regulation for that particular element.

Interactions: Overabundance of one trace element can interfere with the metabolic use of another element available at normal levels. For example, addition of large amounts of zinc to a diet interferes with (antagonizes) intestinal copper absorption, resulting in copper deficiency from a diet with adequate copper content. Copper deficiency can provoke iron deficiency and anaemia. Molybdenum deficiency in animals can be induced by co-administration of large amounts of the similar element tungsten. Iron deficiency can also increase retention of cadmium and lead, and selenium has been proposed to protect against cadmium and mercury toxicity.

3.2   Importance of transition metals in physiological processes

The sheer range of roles of transition metals and the (often complex) mechanisms underlying their chemistry are far beyond the scope of this textbook. However, the following examples provide some insights into some key functional aspects of this important group of elements.

Iron puts the haeme in haemoglobin

Physiologically, iron is the most abundant of the transition metals, reflecting its biological importance. A good illustration of the importance of iron to mammalian life processes lies in the study of the blood's gas transport molecule, haemoglobin. Each red blood cell is packed with around 300 million haemoglobin (Hb) molecules. Indeed, haeme contains iron and the red pigment porphyrin that gives these cells, and blood, the red pigmentation/colouration. In the lungs, oxygen from inhaled air effectively passes into the bloodstream and enters the red blood cells, where it binds haemoglobin (to form oxyhaemoglobin). From here it is shuttled around the circulatory system and is delivered to body tissues to facilitate aerobic metabolism. At the tissues, oxygen is efficiently exchanged for the major metabolic waste product, carbon dioxide (to form carboxyhaemoglobin), and the venous portion of the circulatory system takes this carbon dioxide bound to the haemoglobin back to the lungs where it is removed from the body by exhalation.

Haemoglobin is comprised of four globin (two alpha and two beta) chains, each of which contains a small haeme group (Figure 3.10a). Iron lies at the heart of each of the four haeme groups (Figure 3.10b) and this iron (Fe(II)) effectively provides a lone electron pair acceptor site to bind oxygen (O_2) or carbon dioxide (CO_2). Notably other lone-pair donors such as carbon monoxide (CO) can compete very effectively for the iron electron pair acceptor site. Thus, in an environment rich in gaseous carbon monoxide fumes, carbon monoxide takes up much needed oxygen binding sites, resulting in dangerously inadequate oxygen provision to metabolizing tissues, which can be lethal.

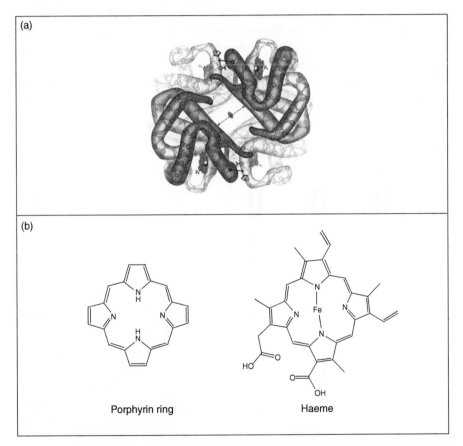

(a)

(b)

Porphyrin ring   Haeme

Figure 3.10 Structural features of haemoglobin. (Illustration, Irving Geiss. Rights owned by Howard Hughes Medical Institute. Reproduction by permission only. Figure 7.5a, page 188, Fundamentals of Biochemistry 2nd edition, Voet, Voet and Pratt.)

The study of simpler synthetic model analogues coupled with the study of dysfunctional forms of protein molecules goes some way to helping understand structural and functional attributes of complex molecules such as haemoglobin. Considering haemoglobin, one of the most important examples is the genetic sickle cell (HbS) mutation of the beta chains of the molecule. The HbS mutation results in production of an aberrant protein, which contains the amino acid valine in place of the normal glutamic acid, near the N-terminus of the beta chain. This seemingly small change to the globin chain results in a change in the electrophoretic migration properties of the haemoglobin protein, allowing bioanalytical identification of HbS versus normal HbA by gel electrophoresis (Figure 3.11a). The physiological consequences of this mutation are more easily understood when observing red blood cells from persons with sickle cell disease compared with normal red blood cells. As shown in Figure 3.11b, the HbS mutation causes the red blood cells to take on an abnormal sickle-like shape (hence the name of the disease). This is as a consequence of the hydrophobic nature of the valine side chain, which results in an aggregation of haemoglobin molecules within the red blood cell and the notable change in shape of the cell.

While simple blood smears can help with diagnosis of HbS, abnormal haemoglobins resulting in other haemoglobinopathies can be investigated by comparing

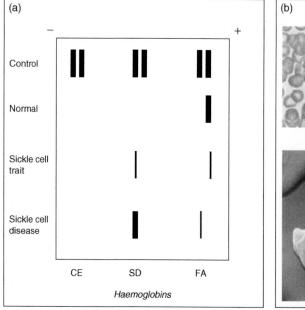
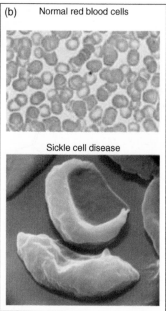

Figure 3.11 **Investigation of abnormal haemoglobin (HbS). Part B figures reproduced from the Encyclopaedia Britannica online (www.britannica.com).**

with HbA. Demonstrable changes include: abnormal electrophoresis (as indicated in Figure 3.11a); ease of denaturation by alkali; changes in oxygen dissociation curve; and alterations in peptide chain structure. There are multiple bioanalytical tools to assess altered peptide composition including mass spectrometry, which is considered in more detail later in this book.

Vitamin B_{12}, the only metal-containing vitamin

As the name suggests, cyanocobalamin (vitamin B_{12}) is a compound of cobalt (Co(III)). With a complex organic structure, this essential water-soluble vitamin is obtained from dietary animal sources and is required for deoxyribonucleic acid (DNA) synthesis, where enzymes that use vitamin B_{12} are involved in the transfer of one-carbon units. The absorption of this vitamin from the gastrointestinal tract only occurs when intrinsic factor glycoprotein is present. While the body can store up to a 12-month supply of vitamin B_{12}, rapid growth or conditions causing rapid cell turnover can increase the body's requirement for this vitamin.

Absorption of vitamin B_{12} can be investigated in several ways after oral delivery of radioactive vitamin B_{12} (e.g. containing ^{58}Co) and subsequent measurement of radioactivity in faecal excretion, whole body counting or liver uptake, plasma radioactivity or the popular Schilling test. For the Schilling test, urinary excretion of radioactive vitamin B_{12} is measured 24 h following oral delivery, and impaired absorption may indicate: intrinsic factor deficiency; bacterial colonization of the small intestine (stagnant gut syndrome); or ileal disease.

The *corrin* ring is the cobalt-binding molecule in vitamin B_{12} (Figure 3.12). With regards to vitamin B_{12} chemistry, much can be learned through the study of simpler analogues.

Methylcobalamin is an important structural derivative of vitamin B_{12} containing a methyl group in place of the cyano group. This organometallic compound, which can transfer methyl groups to metals including gold (Au(I)), mercury (Hg(II)) and thallium (Tl(II)), is involved in the metabolism of methane-producing bacteria. Interestingly, these bacteria are likely candidates for the transformation of elemental mercury into extremely toxic methylmercury species and, as such, an indirect way to study vitamin B_{12} chemistry is through examination of the features and chemistry of methylmercury species.

Vitamin B_{12} is a controlling factor for pernicious anaemia, but since it is difficult to design a diet lacking this vitamin, this condition is very rare. As animal food products are the richest source of vitamin B_{12}, most people that develop deficiency of this vitamin are true vegetarians or infants of vegetarian mothers. Symptoms of B_{12} deficiency have a slow onset, as minute amounts are required and it is efficiently stored, but they include anaemia, a smooth red painful tongue (glossitis)

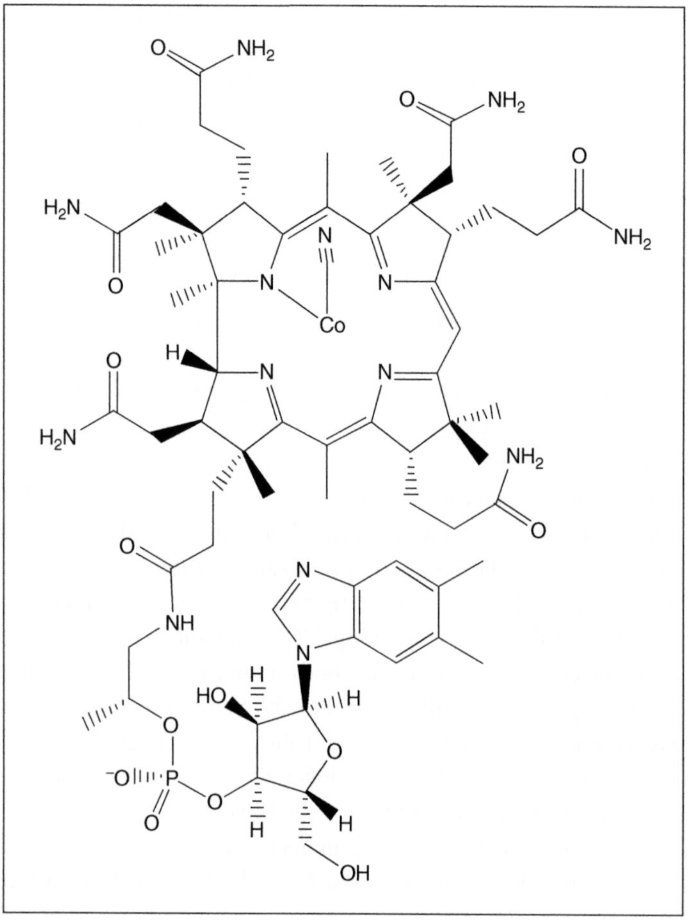

Figure 3.12   **Structure of vitamin B$_{12}$.**

and neurological abnormalities. Conversely, serum vitamin B$_{12}$ is increased in conditions such as chronic obstructive pulmonary disease, congestive heart failure and diabetes.

3.3   Transition metals as mediators of disease processes

While physiological mechanisms usually regulate the amounts of essential elements in the body, for many other elements there are no such control mechanisms, and the amounts in the body reflect the occurrence in food or water. The amounts of a given element in the environment vary, and industrial mining and other human activities can result in the release of main group and transition metals into the environment and ultimately food and water supplies. This can have serious consequences when considering toxic metals such as cadmium, mercury or lead, and also transition metals.

The latter part of the twentieth century saw considerable improvements in the methods used to assess the total amounts and different chemical forms in which a given element may be present in tissue samples or body fluids. These developments also demonstrated the important roles that transition metals play in health and disease, and how accidental or deliberate intake of much larger amounts than in the normal diet can result in the signs, or symptoms of acute or chronic poisoning. For example, acute iron poisoning can result from the ingestion of pharmaceutical products designed to treat iron deficiency, which can have severe gastrointestinal effects (vomiting and diarrhoea) and liver and/or kidney damage in cases where large amounts of iron enter the blood.

The presence of complexing ligands in the food or drinking water may adversely affect the bioavailability of a given transition metal in blood or tissues. For example, some washing powders contain powerful chelating agents such as ethylenediaminetetraacetic acid (EDTA) anions, which may ultimately reach high levels in rivers and drinking water.

As absorption from the gastrointestinal tract regulates the amount of any given transition metal in the blood and tissues, it is perhaps not surprising that genetic or other alterations in metal absorption can result in a deficiency or accumulation of that particular element in the body. However, it is important to note that increased tissue concentrations of metal(s) in a disease state does not necessarily mean that the metal itself actually caused the disorder. For example, while aluminium has been observed in the so-called neurofibrillary tangles (plaques) in the brain that are associated with Alzheimer's disease (premature senile dementia), it is now widely considered unlikely that aluminium causes this neurological condition. Also, the observation that in parts of the world with high concentrations of manganese in the soil there is an increased incidence of amyotrophic lateral sclerosis (ALS; a motor neuron disease) led to the speculation that this disorder may directly result from manganese accumulation. Although this now appears unlikely, multi-element analysis does show differences between normal and ALS tissues, but again this does not tell us whether altered elemental patterns cause the disease or simply arose as a result of ALS-induced alterations to biochemical or physiological pathways regulating one or more elements. Table 3.1 illustrates disorders that have been associated with transition metals.

Table 3.1   **Some disorders which have been associated with transition metals**

Transition metal	Conditions associated with reduced availability or deficiency	Conditions associated with excess availability or accumulation
Chromium	Altered glucose metabolism Glucose intolerance and malnutrition	Liver and kidney impairment Dermatitis Convulsions and coma Respiratory tract and lung cancers (Cr(VI))
Cobalt	Anaemia	Cardiomyopathy/cardiac failure Respiratory sensitization and asthma Allergic dermatitis
Copper	Anaemia Menkes' (kinky hair) syndrome Cardiac abnormalities/heart disease	Wilson's disease Hepatic injury and jaundice Headache, vomiting Haemolytic shock
Iron	Anaemia Nephrosis and uraemia	Siderosis Haemochromatosis Liver cirrhosis Hepatitis
Manganese	Skeletal deformities Gonadal and reproductive dysfunction Defective cholesterol metabolism	Brain abnormalities Respiratory illness Ataxia Motor neuron diseases

Iron deficiency anaemia

Cause/Incidence: Insufficient dietary intake or absorption of iron resulting in problems with haemoglobin production. This condition may also result from physiological changes (e.g. growth spurts, pregnancy) or another underlying pathology such as bleeding from the gastrointestinal tract. Iron deficiency anaemia is the most common form of anaemia encountered globally.

Pathology: Symptoms of this condition arise due to the impact of iron deficiency on haemoglobin synthesis resulting in smaller (microcytic) red blood cells, containing smaller amounts of haemoglobin (hypochromic). These changes can be detected through laboratory investigations, and notably a reduced haemoglobin concentration is a late feature of the condition. The body is able to tolerate low haemoglobin concentrations in this form of anaemia, and thus often there are few symptoms and the disease can go unrecognized for a time. However, pallor, fatigue and general weakness characterize this condition, and severe cases may also result in shortage of breath (dyspnea). These and other symptoms often arise due to the poor oxygen transport around the body.

Treatment: Effective management of this condition relies on determination and treatment of the underlying cause. This may mean treating another pathology (e.g. gastrointestinal blood loss) or iron replacement therapy. For the latter, oral supplements comprising iron(II) (ferrous) salts, and in particular ferrous sulfate, are widely used. Notably, however, chelated iron has an enhanced bioavailability over ferrous sulfate, with the additional advantage that there is no side effect from the sulfur content of ferrous iron. If a cause of iron deficiency is malabsorption, then it may be given parenterally as injected intravenous preparations such as iron dextran or iron hydroxide sucrose. However, parenteral iron is typically not as well tolerated as oral delivery.

Menkes' syndrome

Cause/Incidence: Genetic disorder, first described by Menkes in the early 1960s, resulting in problems with copper distribution through the body. The genetic defect is X-linked recessive, meaning that males (XY) inherit the defect and are affected, while females (XX) may be affected but are more likely to be carriers. While this condition can be inherited, as much as one-third of cases may arise as a result of a new gene mutation. Estimated incidence is somewhere between 1 in 30 000 and 1 in 250 000 individuals.

Pathology: As a result of this genetic disorder, arising from mutations in the so-called *ATP7A* gene (which encodes a copper transport protein), effective copper distribution around the body is severely compromised and onset of the condition typically begins during infancy. Copper accumulates in some tissues (including the small intestine and kidneys), while other tissues (including brain) can have extremely low levels of copper. As copper affects the activity of numerous enzymes (particularly copper-containing enzymes), decreased copper levels may result in impaired cell structure and function, commonly bone, skin, hair, blood vessels, and various tissues of the nervous system. Besides the characteristic kinky, colourless/steel-coloured easily broken hair (hence alternative names kinky hair disease or Menkes kinky hair syndrome), other signs and symptoms include metaphyseal widening, weak muscle tone (hypotonia), seizures, mental retardation and slowed development, and failure to thrive. Unfortunately, the prognosis (disease progression/chance of recovery) is poor, and most individuals will die within the first decade of life.

Treatment: As this condition is usually diagnosed in infancy, some merit can be gained from early treatment with injections (subcutaneous or intravenous) of copper supplements (acetate salts). As the condition develops additional forms of treatment may help directly address other symptoms, however, as noted earlier these merely support the individual during their reduced lifespan.

Classic examples of conditions associated with transition metal storage

Arguably the two most often cited examples of metal storage diseases are Wilson's disease (excess copper) and haemochromatosis (excess iron), which are each considered briefly in turn.

Wilson's disease

Cause/Incidence: Hereditary autosomal recessive genetic disorder first described by Wilson in 1912, which gives rise to altered copper transport resulting in excess metal storage. This condition is more commonly observed in males, with an incidence of around 1 in 30 000, and it is believed 1 in 100 individuals are unaffected carriers.

Pathology: The principal feature of this condition is copper accumulation/storage in liver, brain and eyes causing progressive liver, kidney and brain damage. Despite the fact that the biochemical defects arising from Wilson's disease are present from birth, symptoms usually only appear in late-teens/early twenties and usually the hepatic symptoms appear before the neurological/neuropsychiatric problems. The most common liver (hepatic) disorder is chronic active hepatitis resulting in cirrhosis, but fulminant liver failure (characterized by low alkaline phosphatase and high bilirubin levels) can also occur. Over 80% of patients present with suppressed ceruloplasmin levels, but a more accurate indicator is direct testing of copper levels in a 24 h specimen of urine, in a blood sample or in a liver biopsy tissue sample. An eye examination may also reveal copper deposition in Descemet's membrane of the cornea, giving rise to brown rings (Kayser–Fleisher rings).

Treatment: While damage cannot be cured, disease progression can be slowed down by lifelong chelation therapy. These chelating agents (including D-penicillamine or trientine hydrochloride) can help remove copper from tissue, and zinc supplements may help slow copper absorption, but patients also need to follow a diet low in copper. In extreme conditions where patients do not respond to treatment, liver transplantation may be an option.

Haemochromatosis

Cause/Incidence: Results from improper iron processing/accumulation and chronic iron poisoning, and is the main iron overload disorder. There are two major types of iron overload disorder, categorized under primary and secondary iron overload. The primary forms have genetic origins (spontaneous) while secondary iron overload typically arises due to medical administration (e.g. patients needing frequent blood transfusions to combat severe anaemia are unable to excrete the large amounts of iron that are released from the breakdown of transfused red blood cells). Around 90% of haemochromatosis is associated with mutations in the HFE gene and is often referred to as a *Type 1 primary iron overload*.

Pathology: Four main types (Types 1–4) of haemochromatosis have been described, initial symptoms of which are similar to a wide range of other disorders. As the HFE mutations do not reduce the absorption of iron in response to the higher levels in the body there is a progressive increase in iron stores. As this happens iron is initially stored as ferritin (which can be shed into the blood) or later broken down to form haemosiderin, which can be toxic. Computed tomography (CT) or magnetic resonance imaging (MRI) scans can reveal iron deposits, as can liver biopsies, although the first step in diagnosis is measurement of ferritin in blood along with other markers of iron metabolism (e.g. decreased transferrin and increased serum ion), and if these biochemical markers are altered genetic testing can be conducted.

Treatment: Some of the excess iron can be removed by chelation therapy (e.g. deferoxamine) but if there is early diagnosis, regular phlebotomies (blood letting) can reduce ferritin from very high blood levels (reaching 300 ng l^{-1}) to 50 ng l^{-1}. Other treatments aim to reduce organ damage, and patients should limit intake of alcoholic beverages (associated with increased iron accumulation), vitamin C (which promotes iron absorption), and iron-rich foods (such as red meat). Increased intake of substances that inhibit iron absorption (e.g. high-tannin tea) may also be helpful.

Removal of excess transition metals by chelation therapy

Large ligands can attach to a transition metal ion by two or more bonds, and the ligands may be bi-, tri-, hexa- or polydentate (for 2, 3, 6 or many points of attachment), forming 'chelate rings'. If one or more ligands are competing for a transition metal, typically those offering the more points of attachment are at a significant advantage. That is, if a monodentate ligand competes with a bidentate ligand for a transition metal, the bidentate ligand complex will predominate. Following this principle through, polydentate ligands have the greatest advantage and will be formed to the almost total exclusion of other complexes, hence their widespread use as powerful, effective and specific chelation drugs in transition metal detoxication therapy.

Examples of classic therapeutic chelating agents

Deferoxamine (desferal)

Donor atoms: several O

Removes: $Fe(III)$

EDTA

Donor atoms: 4O, 2N

Removes: $Co(II)$

D-Penicillamine

Donor atoms: S, N, O

Removes: $Cu(I)$, $Cu(II)$

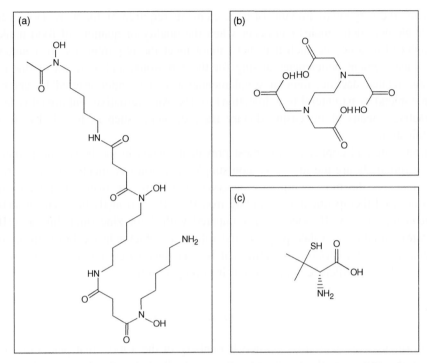

Figure 3.13 **Structures of (a) Deferoxamine, (b) EDTA and (c) d-penicillamine.**

The link between specific transition metals and various disease processes has prompted research interest into new and more effective bioanalytical methods to determine transition metals in physiological fluids. Traditionally clinical chemistry laboratories have used atomic absorption spectrophotometric (AAS) techniques to determine transition metals in physiological fluids. Although both low and high serum levels of various transition metals such as copper, iron and also zinc have been associated with various disease processes, the exact mechanisms underlying many of these conditions are still not completely understood. The following subsections describe some of the most prominent and best understood conditions associated with increased or decreased blood plasma levels of transition metals.

3.4 Therapeutic implications of transition metals

The recommended daily amount of many trace metals varies widely and is controlled by a variety of factors including growth, age, general fitness and diet. While virtually impossible to establish the exact requirements for daily dietary intake of all nutrients, professional bodies have tabulated average recommended values for many elements including transition metals. In most cases the diet can provide the very small amounts of trace elements required by the body. However, trace element deficiencies can occur when the quality or quantity of food intake is reduced to a degree such that diet cannot meet the requirements. To counteract trace element deficiencies arising for these reasons, and as result of disease or age-related decline, mineral supplementation can be administered by enteral (e.g. tablets) or parenteral (e.g. injection) routes. An alternative, but arguably less attractive, means of increasing dietary trace elements such as iron is by 'food fortification'.

Iron, zinc and copper are the most prevalent metals in the body, and perhaps not surprisingly are the most common supplementation treatments. Of these, iron deficiency is the most commonly observed and there are over 40 preparations of iron used therapeutically (including iron(II) sulfate; iron(II) fumarate; iron(II) gluconate; and iron(II) succinate), compared with 3 for zinc (including zinc(II) sulfate). Notably, as folate plays a vital role as a cofactor in the biosynthesis of haeme (for haemoglobin), sometimes it is necessary to co-administer iron with folic acid in so-called 'compound oral iron preparations'.

Transition metal chelation and anticancer therapies

The primary observation by Barnett Rosenberg in the 1960s, that platinum wire could affect the growth of cultured cells, eventually led to the discovery that certain coordination compounds of group VIII transition metals may have cytotoxic anticancer properties. The first of these agents, cisplatin (*cis*-dichlorodiammine platinum(II)), introduced therapeutically in the 1970s, was demonstrated to form a platinum chelate with bases of the DNA molecule (intrastrand link), effectively interfering with DNA replication and thus cell division. The cytotoxic (cell-killing) anticancer effect lies in the reduced ability of tumour cells to repair these cisplatin-induced DNA intrastrand breaks. The clinical efficacy of intravenously infused cisplatin, used alone or in combination with other cytotoxic drugs against ovarian, testicular and lung cancer, prompted much research into additional transition metal-based anticancer therapeutics. These have included second generation drug carboplatin and the orally active platinum drug satraplatinum (see Figure 3.14).

Radioprotective properties of copper complexes

Many free-radical scavengers (including dithiols and dithiocarbamates) have potential therapeutic usefulness as radioprotective agents. Copper complexes are known to be scavenging agents for the superoxide radical, which is believed to play a role in the induction of radiation damage. The toxic effects of superoxide are believed to lie in its ability to reduce metal ions, for example Cu(II) to Cu(I),

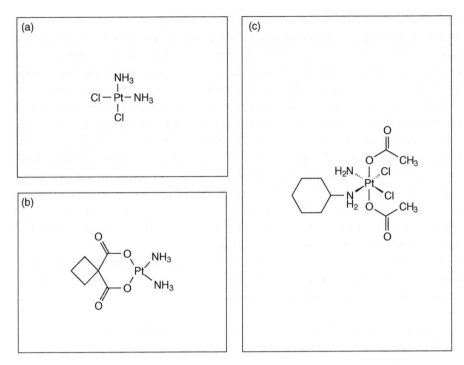

Figure 3.14 **Structures of (a) cisplatin, (b) carboplatin and (c) satraplatinum.**

forming OH• radicals that combine to form cytotoxic hydrogen peroxide (H_2O_2). However, copper(II)di-isopropylsalicylate (Cu-DIPS) appears to effectively reduce radiation damage long after the radiation-induced free radicals would have disappeared, suggesting other beneficial actions of copper in cytoprotection and tissue repair.

3.5 Determination of transition metals in nature

Transition metal complexes have special properties, including colour and unusual magnetic properties that are useful in bioanalysis. Traditionally the most widespread technique for determining transition metals in physiological and biological fluids has been atomic absorption spectrophotometry.

Atomic absorption spectrophotometry

There are two major analytical techniques to determine the concentration of particular metal elements in samples, namely *atomic absorption spectroscopy* and *atomic absorption spectrophotometry*. It is important to recognize that these two analytical techniques are not identical. While atomic absorption spectroscopy is more widely used and can determine the concentration of over 62 different metals in solution, this method is not as accurate as atomic absorption spectrophotometry (AA or AAS), which is the preferred method for determination of transition metals in solution. Indeed, AAS can detect very low concentrations of transition metals (less than one part per million) in small samples, representing a very sensitive form of spectrophotometry. While spectrophotometry will be covered in more detail in Chapter 5, briefly, an atomic absorption spectrophotometer turns the sample solution into an aerosol before atomization using an intense flame. Light emitting from a hollow cathode lamp (specific to the metal being detected) passes through the atomized sample, and a photomultiplier tube detects the light absorbed by the metal in the sample. This is recorded by the atomic absorption spectrophotometer and output as a data reading.

As mentioned above, the different colorimetric qualities can give some information as to the transition metal, and the following simple examples of different colourations of transition metal-containing compounds:

Chromium(II) halides: CrF_2 is green; $CrCl_2$ is white, $CrBr_2$ is white, and CrI_2 is red-brown.

Chromium(III) halides: CrF_3 is green; $CrCl_3$ is red-violet; $CrBr_3$ is green-black; and CrI_3 is black.

Chromium oxides: CrO_3 is deep red; CrO_2 is brown-black; Cr_2O_3 is green.

Iron(II) halides: FeF_2 is white; $FeCl_2$ is pale yellow-grey; $FeBr_2$ is yellow-green; and FeI_2 is grey.

Iron(III) halides: FeF_3 is green; $FeCl_3$ is black; $FeBr_3$ is dark red-brown.

Iron oxides: FeO is black; Fe_2O_3 is red-brown.

Manganese(II) halides: MnF_2 is pale pink; $MnCl_2$ is pink; $MnBr_2$ is rose; and MnI_2 is pink.

Manganese oxides: MnO is grey-green; MnO_2 is black; Mn_2O_3 is black.

Limitations in colorimetric and atomic absorption spectrophotometric measures have prompted development of alternative methods of transition metal analysis. One example is the use of ion exchange chromatography to assess transition metals in serum and whole blood, a method developed by the company Dionex. While identification of transition metal complexes can also be made on the basis of symmetry (Laporte rule) or spin selection rule or analysis of charge transfer spectra, one of the most significant methods is on the basis of magnetism. The

Table 3.2 Sillén's estimates of the impact of mining on environmental distribution of transition metals

Element	Geological dissolution by rivers (kg per annum)	Additions from mining (kg per annum)
Copper	3.9×10^8	4.6×10^9
Iron	2.5×10^{10}	3.3×10^{11}
Manganese	4.5×10^8	1.6×10^9
Molybdenum	1.3×10^8	5.8×10^7

principles of magnetism are discussed in Chapter 11 (nuclear magnetic resonance; NMR) and techniques based on NMR can be used for analysis of transition metals.

Mining for minerals has inevitably released metals contained in rocks into the environment. In the 1960s, Sillén estimated the additional amounts of metals entering into the environment and highlighted the necessity for consideration before disposing of metal-containing waste into the environment (Table 3.2).

Key Points

- A transition metal is defined as an element, an atom of which contains an incomplete d shell, or that gives rise to a cation with an incomplete d shell.

- The coordination number denotes the number of donor atoms associated with the central atom and dictates the shape (stereochemistry) of the coordination complex.

- Oxidation describes the process by which electrons are removed from the metal, whereas reduction describes the process by which electrons are added to the metal.

- Oxidation state refers to the number of electrons added to or subtracted from the metal, and transition metals can adopt a number of oxidation states.

- Transition metals are integrated into complex structures consisting of a central metal ion or atom surrounded by a set of ions or molecules called *ligands*.

- Key reactions include (a) substitution; (b) electron transfer; (c) activation of ligands; (d) redox-catalyzed substitution; (e) oxidative-addition reaction; (f) insertion; (g) redox-catalyzed insertion.

- Transition metals play important roles in health and disease and intake of larger amounts than in the normal diet can result in the signs/symptoms of poisoning.

- Important disorders associated with transition metals include insufficient intake or absorption (e.g. iron deficiency anaemia), defective transport (e.g. Menkes' syndrome), and excessive storage (e.g. Wilson's disease, haemochromatosis).

- Excess transition metals can be removed by chelation therapy using chelating agents such as deferoxamine, EDTA or D-penicillamine, and supplements can be used in cases of deficiency (e.g. iron(II) suphate or zinc(II) sulfate).

- Transition metal complexes have special properties including colour and unusual magnetic properties, and atomic absorption spectrophotometry is the most widespread technique used to determine transition metals in biological fluids.

4 Ions, electrodes and biosensors

Ions play a key role in the regulation of physiological and chemical processes. As such, it is essential to be able to monitor and measure various different ions and ionic compounds in biological and environmental samples. This branch of bioanalytical chemistry spans a number of core technologies that are generally categorized under the umbrella terms *sensors* and *biosensors*. Historically, this area has its beginnings in the early 1800s, when the British chemist/physicist, Michael Faraday, proposed that biomolecules contained 'ions' which acted as the key chemical entities mediating electrical current. Faraday is widely considered to have been one of the finest scientists ever, and other relevant contributions to this field include description of the laws of electrolysis, and popularization of the terms anode, cathode, and electrode. Another legend in this field is the Swedish physical chemist Svante August Arrhenius, who made major steps in formulating mechanisms underlying Faraday's earlier proposal. While Faraday hypothesized that ions were produced during electrolysis, Arrhenius suggested that salts in an aqueous solution (now termed *electrolytes*) dissociated into these charged particles. Arrhenius received the Nobel Prize in Chemistry in 1903 for his 'electrolytic theory of dissociation', described in his doctoral thesis. The history of chemical and biochemical sensors goes back to the mid 1800s when Siemens built the first sensor, later patenting his invention (an electromechanical transducer). Interestingly, this entrepreneurial family founded and led the German company known today as *Siemens AG*, one of the largest global electrotechnological companies. Arnold Beckman is considered to have been the father of commercial biosensors, and was founder of a company known today as *Beckman Coulter Inc.*, which he established to commercialize his development of modern glass electrodes, in particular for pH measurement. This company is still one of the world leaders in

Understanding Bioanalytical Chemistry: Principles and applications Victor A. Gault and Neville H. McClenaghan
© 2009 John Wiley & Sons, Ltd

the development of innovative bioanalytical instruments. The following sections provide a concise overview of ions, electrodes and biosensors with key applications in life and health sciences.

Learning Objectives

- To relate knowledge of the impact of ions and oxidation-reduction reactions on physical and life processes.

- To understand the importance of pH and biochemical buffers in physiological regulation and be able to perform related calculations.

- To explain core features and applications of different chemical and physical sensors and biosensors.

- To outline and convey knowledge of how specific electrodes are used to measure various parameters.

- To give specific examples of biosensors and their applications in life and health sciences.

4.1 Impact of ions and oxidation–reduction reactions on physical and life processes

Most biomolecules in living organisms exist in a charged state attributed to their containing ions. Fundamentally, an ion is an atom (or group of atoms) that has lost or gained one or more electrons through oxidation or reduction. These terms are extremely important to understand as they lie at the heart of many physiological and chemical changes regulating life processes. Good examples include the electron-transport chain (respiration) and numerous fundamental metabolic reactions.

Electrons: loss and gain

Ionization is a term used to describe the loss or removal of one or more electrons from an atom. This process requires application of external energy to the atom, which is called the *ionization energy*. As the name suggests, the process of ionization is the means by which ions are formed. This phenomenon, first described by Arrhenius, is key to understanding oxidation and reduction and the classification of acids and bases, which are considered briefly in the following.

Oxidation and reduction

As with most concepts involving electrons, oxidation and reduction reactions are often initially misinterpreted as complicated and difficult to understand. Oxidation and reduction are simply complementary processes involving the loss and gain of electrons from molecules, atoms or ions. Whereas oxidation is the loss of one or more electrons (i.e. oxidation is loss (OIL)), reduction is gain of one or more electrons (i.e. reduction is gain (RIG)). These abbreviations are an easy way to remember the difference between these two processes with respect to electron changes (OIL RIG). As these processes are complementary and occur in the same system they are often referred to as *redox* reactions (i.e. *red*uction and *ox*idation). Figure 4.1 provides a simple illustration of this principle.

Importantly, the terms *oxidation* and *reduction* can also refer to the loss or gain of hydrogen atoms or oxygen atoms from molecules. In this case, oxidation refers to the loss of hydrogen atoms *or* gain of oxygen atoms and reduction is the gain of hydrogen atoms *or* loss of oxygen atoms.

Substances that can oxidize or reduce other substances are known as *oxidizing agents* (oxidant) or *reducing agents* (reductant), respectively. An oxidizing agent removes electrons from another substance (i.e. causes oxidation), thereby gaining electrons (i.e. is an electron acceptor), and as reduction is gain (RIG), the oxidizing agent itself is reduced. Conversely, a reducing agent gains electrons from another substance (i.e. causes reduction), thereby losing electrons (i.e. is an electron donor), and as oxidation is loss (OIL), the reducing agent itself is oxidized. In reality, these simplistic descriptions are not entirely accurate, as oxidation and reduction do not necessarily describe the actual physical transfer of electrons, so it is more correct to consider oxidation as an increase in oxidation number and reduction as a decrease in oxidation number.

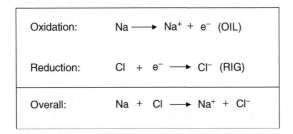

Figure 4.1 **Example of a simple oxidation–reduction reaction.**

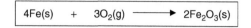

$$4Fe(s) \quad + \quad 3O_2(g) \longrightarrow 2Fe_2O_3(s)$$

Figure 4.2 **Chemical equation for the formation of rust.**

Processes regulated by oxidation and/or reduction

The process of corrosion and electrochemical reactions are very visible and useful illustrations of the oxidation–reduction process. As metals are exposed to oxygen in air and water, the constituent metal atoms become oxidized, forming metal ions/metal oxides. Put simply, the oxidation–reduction process results in the metal rusting and becoming severely corroded and weakened. In the case of environmental exposure of iron, the metal forms iron(III) oxide, known as *rust* (see Figure 4.2).

Also, statues with copper skins are particularly prone to the effects of oxidation–reduction processes, and good examples are the Statue of Liberty or roofs of old buildings. While these would have had a characteristic rich copper-brown coloration, environmental exposure caused oxidation of the copper and other constituent metals, changing their appearance to a light-green coloration with additional corrosion.

While there are very many good examples of oxidation–reduction processes in nature, perhaps the two most classic examples are aerobic respiration and alcohol metabolism by the liver, each of which is considered briefly in turn here.

Aerobic respiration: The mitochondrial electron-transport chain incorporates cytochrome c, and the oxidation and reduction of the iron atoms in this biomolecule passes electrons through a series or chain of metabolic reactions. This is illustrated in Figure 4.3.

Alcohol metabolism: Liver enzymes play a key role in the metabolism of alcohol (ethanol). As the liver oxidizes ethanol it generates acetaldehyde – the substance which helps create the hangover. The acetaldehyde is then further oxidized to acetic acid and eventually CO_2 and H_2O.

Notably, oxidation–reduction processes and so-called redox reactions are the basis of the workings of electrochemical cells considered in more detail below.

Ions and electrochemical cells

An ion with a positive charge is known as a *cation* while an ion with a negative charge is termed an *anion*. Ions have unequal numbers of electrons and protons and an atom of, for example, hydrogen which loses an electron becomes a cation with a positive charge, denoted as $+1$; this is the hydrogen ion (H^+) which is the simplest

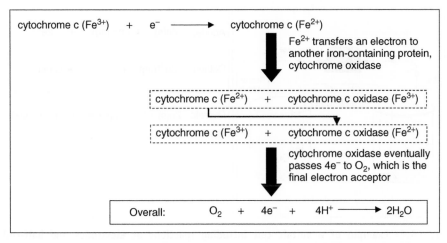

Figure 4.3   **Example of oxidation–reduction reactions in biological respiration.**

ion, and arguably one of the most important ions in life processes. However, it is important to note that ions can comprise a single atom (monatomic ion) or multiple atoms (polyatomic ion), and ions with many atoms are often referred to as molecular ions. Ionic bonds are formed by the electrical attraction between anions and cations, and it is normally assumed that negatively charged anions (e.g. Cl^-) are attracted to the positive anode and repelled by the negative cathode. However, this concept causes considerable confusion for several principal reasons: (i) on the one hand *an*ion refers to a negative charge, while *an*ode refers to a positive electrode (see Figure 4.4), but (ii) as discussed below, although the anode is the positive electrode in an electrolytic cell, the opposite is true in a voltaic cell (where the anode spontaneously acts as the negative electrode). In order to help clarify these issues, voltaic and electrolytic electrode systems (so-called *cells*) are discussed below, with particular emphasis on inherent differences between each type of cell.

Voltaic (Galvanic) cell

The voltaic cell is credited to the Italian physicist Alessandro Giuseppe Antonio Anastasio Volta and his invention of the voltaic pile in the 1800s, the first modern battery comprising zinc and copper electrodes. While this is considered the first demonstration that when metals and chemicals come into contact they generate an electrical current, there was some professional friction between Volta and another Italian scientist Luigi Galvani. This related to the prior observation by Galvani in the late 1700s, that when two different metals were connected together and simultaneously touched different parts of a nerve in a frog's leg it caused a muscle

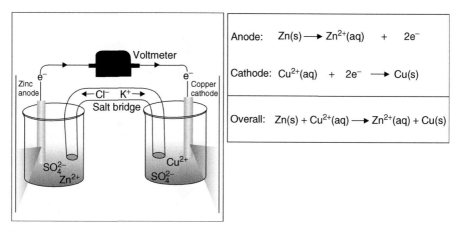

Figure 4.4   Example of a voltaic cell including reactions at electrodes and half-cell equations.

twitch (contraction). Although the discovery of bioelectricity and galvanic cell is attributed to Galvani, this term is now interchangeable with the arguably more popular name voltaic cell, after Volta. The so-called voltaic cell comprises two half-cells (that is, two independent containers) connected by a tube filled with an electrolyte (salt bridge) that converts inherent stored chemical energy into electrical energy, which can be recorded on a voltmeter. An electrolyte is loosely described as a material that dissolves in water to produce a solution which is able to conduct an electrical current. While electrolytes can be acids, bases or salts, the latter are important in electrochemistry, and sodium chloride and potassium chloride are strong electrolytes. In a voltaic cell, an oxidation reaction occurs at one half-cell (the anode, 'electron donor'), while a reduction reaction occurs at the other half-cell (the cathode, 'electron acceptor'). For example, a voltaic cell could have a zinc anode immersed in an aqueous solution of $ZnSO_4$ comprising one half-cell, and a copper cathode immersed in an aqueous solution of $CuSO_4$ comprising the other half-cell. Importantly, both half-cells are connected by a salt bridge, usually KCl, which allows completion of an electric circuit. When the reactive Zn metal electrode is dipped into the $ZnSO_4$ solution there is a spontaneous chemical oxidation reaction where zinc atoms (Zn) in the electrode are oxidized to zinc ions (Zn^{2+}), releasing electrons. At the top of the anode half-cell lying outside the solution is a copper wire directly connected to a copper rod (the cathode) in the other half-cell. Electrons generated from the oxidation reaction at the anode flow through the copper wire and voltmeter to the cathode, as illustrated in Figure 4.4. When electrons arrive at the copper rod they cause an electrochemical reaction whereby reduced copper ions (Cu^{2+}) from the $CuSO_4$ solution form copper atoms (Cu) at the cathode. To complete the electric circuit, current flows from the cathode to the anode through the salt bridge wherein Cl^- ions (negatively charged) are attracted towards the positively charged anode; see Figure 4.4.

This type of cell essentially operates like a simple battery, with many diverse applications, and it is anticipated that such voltaic cells could be charged by the human body to provide a future power source for implanted medical devices such as heart pacemakers.

Electrolytic cell

To add to the confusion, the charges on the electrodes on the voltaic cell are the reverse of those found in the electrolytic cell. This is due to the application of an external battery, which forces electrons to flow through the circuit in the opposite direction. As such, the electrodes (such as Zn and Cu), electrolytes ($ZnSO_4$ and $CuSO_4$) and salt bridge can be identical in both a voltaic and electrolytic cell, but the systems work in a different manner. Whereas in a voltaic cell spontaneous oxidation and reduction reactions occur, in an electrolytic cell the oxidation or reduction reactions at the electrodes are non-spontaneous, arising as a direct result of, and driven by, the application of external electrical energy. Put simply, in a voltaic cell spontaneous oxidation causes electrons to flow from the Zn electrode (negative anode) towards the Cu electrode (positive cathode) and this system can be converted into an electrolytic cell by simply attaching a battery with a voltage strong enough to drive electron current towards (as opposed to away from) the Zn electrode, thus reversing electron flow through the circuit. As such, in a voltaic cell oxidation occurs at the Zn electrode and in an electrolytic cell oxidation occurs at the Cu electrode. It is critical to note that, by definition, electrons always flow towards the cathode, so as illustrated in Figure 4.5, in a voltaic cell the cathode is the Cu electrode (which is positive) and in an electrolytic cell the cathode is the Zn electrode (which is negative).

By virtue of the fact that the external battery is essentially forcing electrons into the part of the system (half-cell) that will spontaneously release electrons through an oxidation reaction, it is acting to charge the system, by forcing the negative electrode to accept negatively charged electrons. This action ensures that the 'battery' becomes fully recharged, enabling it to spontaneously emit an electron current in the absence of the external electrical charge.

4.2 pH, biochemical buffers and physiological regulation

The basis of many chemical reactions depends on the physical conditions and environment in which the reaction takes place. For example, many enzymes will not function properly if the surrounding fluid is too *acidic* or *alkaline (basic)*.

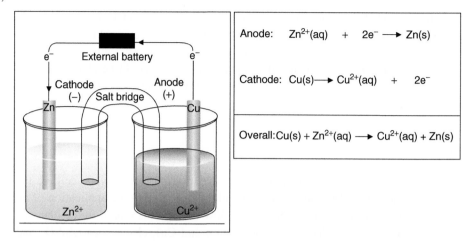

Anode: $Zn^{2+}(aq)$ + $2e^-$ ⟶ $Zn(s)$

Cathode: $Cu(s)$ ⟶ $Cu^{2+}(aq)$ + $2e^-$

Overall: $Cu(s) + Zn^{2+}(aq)$ ⟶ $Cu^{2+}(aq) + Zn(s)$

Figure 4.5   Typical set-up and reactions in a basic electrolytic cell.

As such, acidity, alkalinity and pH are extremely important factors regulating physiological and pathophysiological processes.

Acids and bases

It is important from the outset not to confuse the underlying principles governing acid–base reactions with those of oxidation–reduction reactions outlined above. While they both inherently rely on some sort of charge-transfer process, fundamentally acid–base reactions involve transfer of positive charge (in the form of protons or hydrogen ions) and oxidation–reduction reactions entail transfer of negative charge (in the form of electrons). Characteristically, an acid is a proton donor, dissociating in water to from hydrogen ions or protons, while a base is a proton acceptor, dissociating in water to form hydroxide ions. Classic examples of an acid and base are hydrochloric acid and sodium hydroxide (Figure 4.6).

Notably, as water has both acid and base properties it is termed *amphiprotic* and is the most common solvent for acid–base reactions. There are 'strong acids' and 'weak acids' and 'strong bases' and 'weak bases'. The primary feature determining the so-called strength of an acid or base relates to its degree of dissociation in solution, that is, the fraction that produces ions in solution. Importantly, acid or

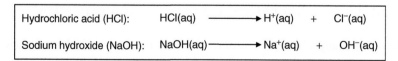

Hydrochloric acid (HCl): $HCl(aq)$ ⟶ $H^+(aq)$ + $Cl^-(aq)$

Sodium hydroxide (NaOH): $NaOH(aq)$ ⟶ $Na^+(aq)$ + $OH^-(aq)$

Figure 4.6   Dissociation equations for HCl and NaOH.

base dissociation in solution does not depend on its relative concentration, that is, a higher concentration of a given acid does not produce more H^+ ions in a given solution. The concentration of a given substance (e.g. X) is abbreviated and denoted as [substance] (e.g. [X]). When the fraction of an acid or base that produces ions in solution is extremely high (almost 100%) then the acid or base is termed *strong*, whereas if the fraction is low it is referred to as 'weak'. Commonly encountered strong acids are hydrochloric acid (HCl) and sulfuric acid (H_2SO_4), and weak acids are acetic acid (CH_3COOH) and carbonic acid (H_2CO_3). Familiar strong bases include sodium hydroxide (NaOH) and potassium hydroxide (KOH), and weak bases include pyridine (C_5H_5N) and methylamine (CH_3NH_2). As noted earlier, aqueous solutions of acids and bases can also behave as electrolytes, where strong acids and bases are strong electrolytes and weak acids and bases are weak electrolytes – where the conductivity of the electrolyte is dependent on the solute (acid/base) and not the solvent (water).

pH and its regulation

Acidity or alkalinity of a solution determines or describes the pH of that solution, following a scale with high pH or low pH, much like temperature may be described as hot or cold. The pH scale, while having no units, is nonetheless not an arbitrary scale, rather it is a reverse logarithmic representation of the relative hydrogen ion concentration described when water (H_2O) combines with hydrogen ions (H^+) to form hydronium ions (H_3O^+) as denoted by the following equation:

$$pH = -\log_{10}[H_3O^+] \quad \text{or arbitrarily} \quad pH = -\log_{10}[H^+]$$

As pH scale reflects proton (H^+) concentration, the relationship is not linear, rather logarithmic, where a shift in value of 1 on the scale equates to a 10-fold difference in H^+ concentration. So, a shift in pH from 2 to 3 is a 10-fold decrease in H^+ concentration, whereas a shift from 7 to 5 represents a 100-fold increase (i.e. 10×10) in H^+ concentration. Solutions with pH values on the scale of less than 7 are acidic (i.e. H_3O^+ predominates), while those with pH values greater than 7 are alkaline/basic (i.e. OH^- predominates). A neutral pH of 7, where H_3O^+ equals OH^-, is attributed to pure water at temperature 25° C (approximately room temperature). So when adding an acid to water the acid acts as a proton donor increasing H_3O^+ and decreasing OH^-, while the converse is true if a base is added to water, and in each case $[H_3O^+] \times [OH^-] = 1 \times 10^{-14}$, a number known as the *ion product*. A pH scale is given in Figure 4.7.

The concept of pH was first described by the Danish chemist Søren Peder Lauritz Sørensen, and while it is not clear where the exact term pH actually came from, there are various suggestions from Latin, French, and English terms. However, for the purposes of this book it is important to remember that it simply

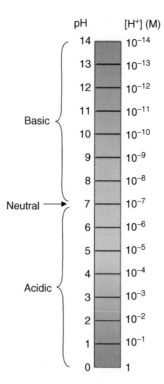

Figure 4.7   Diagrammatic representation of the pH scale. (Adapted from Pratt & Cornely, Essential Biochemistry, 2004.)

describes the outcome of a chemical calculation following the equation noted above, and pH is measured using a particular device referred to as a *pH meter* (see later). Additionally, a few key calculations based on the pH equation are given below to illustrate how pH relates to H^+ concentration in solution.

Calculate the pH of a solution with an H^+ concentration of 1.5×10^{-8} M (remember minus a minus equals a plus).

$$pH = -\log_{10}[H^+] = -\log_{10}[1.5 \times 10^{-8}] = -(-7.82) = 7.82$$

Knowing the pH of a solution is 5.24, calculate the hydrogen ion concentration.

$$pH = -\log_{10}[H^+] \quad \text{so} \quad 5.24 = -\log_{10}[H^+]$$

so inverse $\log_{10}$ of $(-5.24) = [H^+]$ and from this $[H^+] = 5.75 \times 10^{-6}$ M

Calculate the pH of a 1×10^{-6} M solution of the strong base, NaOH.
When considering a base, rather than thinking only about $[H^+]$, focus is placed on $[OH^-]$. From the information given, we need to find a value of $[H^+]$ (or $[H_3O^+]$) to substitute into the pH equation, in this case $[OH^-]$, equals 1×10^{-6} M.
As $pH = -\log_{10}[H_3O^+]$ we first need to calculate the $[H_3O^+]$ concentration of this base.

Remembering that $[H_3O^+] \times [OH^-] = 1 \times 10^{-14}$

then $[H_3O^+] = 1 \times 10^{-14}/[OH^-]$

so $[H_3O^+] = 1 \times 10^{-14}/1 \times 10^{-6} = 1 \times 10^{-8}$

and using this value $pH = -\log_{10}[1 \times 10^{-8}] = 8$

As mentioned earlier, understanding the pH equation and the regulation and control of pH is fundamentally important when considering very many life and health processes. A simple indication of the importance of environmental pH is for growth of crops (soil pH) and acid rain (water pH), which can affect the ecosystem. Indeed, optimum conditions for purification of water and sewage treatment also are pH dependent. Physiologically, pH is critical to maintain normal body functions and key to biochemical reactions in the blood and other body fluids. Buffers and buffer systems are the primary means to regulate and maintain pH, and are discussed in more detail below (with examples in Appendix 3).

Biochemical buffers

Buffers are a means of resisting or modulating changes in pH of solutions in order to keep the environment constant, to allow optimal conditions for biochemical reactions. As such, solutions which can buffer (so-called buffer solutions) act to counter the change in $[H_3O^+]$ and $[OH^-]$, and hence pH, when coming into contact with acid or base or upon dilution. Buffer solutions comprise both a weak acid and conjugate base, or alternatively (and less frequently) a weak base and conjugate acid, where the resistance comes from the equilibrium in the interactions between the weak acid (HA) and conjugate base (A^-).

This equilibrium is illustrated as: $HA + H_2O \rightleftharpoons H_3O^+ + A^-$

The action of buffer systems to resist change follows Le Chatelier's principle, where the system acts to restore the disturbance in equilibrium. So if alkali is added to this system there is a build up on the right-hand side of the equation that shifts the equilibrium to the right, and in order to maintain balance, more acid (HA) needs to be dissociated. Conversely, if strong acid is added to the system there is a build up on the left-hand side of the equation that shifts the equilibrium to the left, and in order to maintain balance, more base (A$^-$) needs to be protonated (i.e. becomes HA). It is important to note that in circumstances where very strong acids or bases are added to a buffer system it may not be able to restore equilibrium and there will be a large associated change in pH.

When considering the behaviour of acids in solution it is important to consider the so-called acid dissociation constant (denoted K_a). Put simply this is a means of describing the dissociation of HA into constituent H$^+$ and A$^-$, which is represented by the following equation:

$$K_a = ([H^+] \times [A^-])/[HA]$$

This can be further manipulated to generate another equation (the Henderson–Hasselbalch equation) relating pH to pK_a (i.e. $- \log_{10} K_a$), which is as follows:

$$pH = pK_a + \log_{10}([A^-]/[HA])$$

where [A$^-$] is taken as the concentration of conjugate base and [HA] is taken as the concentration of the acid, and maximum capacity of a system is when pH $= pK_a$.

Again, buffers are essential in keeping pH within defined limits so that biochemical reactions can occur with maximum efficiency (i.e. maintaining so-called homeostasis). If, for example, the pH gets too low, the acidic environment can damage (or denature) and thus disable enzymes. An important example is the buffering of blood by the carbonic acid–bicarbonate buffer system illustrated below:

$$H_2CO_3 + H_2O \rightleftharpoons H_3O^+ + HCO_3{}^-$$

So the carbonic acid (H$_2$CO$_3$) acts as the weak acid, and the bicarbonate ion (HCO$_3{}^-$) acts as the conjugate base, and this system maintains pH within the range 7.35–7.45.

4.3   Chemical and physical sensors and biosensors

There are many types of *sensor* and *biosensor*, and it is useful to start by considering the key differences between common terms associated with these devices. Useful examples of sensors include the nose (physiological) and a mercury thermometer (direct-indicating), and in essence the word sensor is attributed to a whole device that detects, records, and/or measures a physical/chemical property and may ultimately respond to this property. There are varying definitions and uses of the terms sensor and biosensor, but for the purposes of this book: (i) a *physical sensor* measures physical quantities including electricity, length, temperature and weight; (ii) a *chemical sensor* selectively measures specific chemical substance(s) in a sample in a qualitative or quantitative manner; (iii) a *biosensor* measures a biological component in a sample, and often biosensors are types of chemical sensor which recognize substances which are biological, such as an enzyme, antibody or receptor. In general, by their very nature, chemical sensors/biosensors are usually more technically complex than physical sensors.

Features of sensors

When considering a sensor there are two primary considerations: firstly the ability of an inherent component to selectively recognize a substance in a sample – the so-called *recognition element*, and secondly the ability of the device to detect the substance – the so-called *transducer* (or detector device). A transducer is a device that converts a physical or chemical change into a measurable signal, and as such the transducer is linked to the recognition element (for example, by adsorption). There are four principal types of transducer: (i) *electrochemical* (which includes potentiometric, voltammetric and conductometric); (ii) *optical transducer* (used in various forms of spectroscopy); (iii) *piezo-electric* (where electric currents emerge from vibrating crystal); and (iv) *thermal* (signals generated on the basis of temperature change). The most important features of sensors are their capability, and some key factors determining their performance are given in Table 4.1.

Applications of sensors and biosensors

Sensors are an important aspect of everyday life, and while physiological sensors (e.g. nose and tongue) are taken for granted, other electrical or electronic sensors are encountered in very many industrial and medical devices. Importantly, technological advances have miniaturized sensor devices and these may now even be microscopic. Furthermore, sensors play an important role in diverse research areas encompassing biochemistry, biology, geology and oceanography, and have uses in forensic analyses and clinical diagnostics as well as very many industrial aspects including quality control and environmental monitoring. Some important examples of medical uses of physical and chemical sensors are outlined below.

Table 4.1   Factors determining the performance of sensors

Factor	Notes
Selectivity	Ability of the sensor device to detect only one parameter and/or discriminate between different substances in a sample. Primarily dependent upon recognition element. Arrays are often used to compensate when selectivity is not assured.
Sensitivity	Relationship between the sensor's output and input, so that changes in output correspond to changes in input, often generating a constant.
Span or range	Sensors will have a predefined ability to detect/measure a substance within a given scale with maximum and minimum values, which may reach very low levels (e.g. femtomolar concentrations).
Measurement errors	In all measurement devices it is important to consider issues relating to inherent operating errors in the system. Some important errors relate to: bias (where output is not zero when input is zero); noise (random deviation of signal); drift (slow change in output signal independent of input). Many of these systematic or random errors relate to gain (mean ratio of output to input), and when the measured property lies outside the sensor range values are only estimates.
Accuracy	How closely the value measured on the sensor matches the real or true value. For sensors a deviation of <5% is usually acceptable.
Resolution	Smallest incremental change that can be detected by the instrument on its read-out scale, which relates to precision. Notably some sensor devices can resolve atoms.
Precision	How reproducible the measurements (repeated measurements) are with respect to each other, in other words how closely they match each other. This usually will incorporate statistical analysis of the data and generate error values (e.g. standard error of the mean).
Sample solution	Sensor readings will potentially be affected by deviations or fluctuations in conditions such as pH, ionic strength and temperature of a solution.
Time	Usually inherent to operation of the sensor are response time and recovery time, and lifetime. It may take seconds or minutes for the device to equilibrate before a measure can be taken, thus reflecting response time, and the time interval before a subsequent reading can be taken is the recovery time. Lifetimes of sensors vary and are often dependent on the stability of the recognition element, which may be irreplaceable.
Calibration	Process of determining the relationship between recorded output and input. Mostly this involves adjusting the output to agree with a preset value(s) from one or more standards (e.g. solution with known pH = 4). Where there are multiple standards used this will generate data, which can be plotted graphically as a standard curve or calibration curve.

Physical sensors: (i) Thermal measurement (e.g. core body temperature, surface temperature mapping); (ii) mechanical measurement (e.g. non-invasive sphygmomanometer for blood pressure measurements, spirometer for determination of breathing and pulmonary function); (iii) acoustic measurement (e.g. ultrasound imaging, Doppler sonography for determination of blood flow) and (iv) radiation measurement (e.g. X-ray imaging, CT scanning).

Chemical sensors: (i) Gases (e.g. blood oxygen electrode, carbon monoxide detector); (ii) pH and ions (e.g. pH meter, potassium-selective electrode); and (iii) optical oximetry (e.g. pulse oximetry for non-invasive monitoring of blood oxygenation).

4.4   Important measurements using specific electrodes

Glass electrodes and pH

Glass membrane electrodes are used for measurement of pH (H^+) and Na^+, and may also be used as a transducer for pCO_2 sensors, which are used to measure the partial pressure of carbon dioxide in body fluids. As pH is an important factor regulating chemical reactions, the maintenance of pH within a certain range (between set points) is important for maintenance of physiological function and homeostasis (balance). pH is measured using an instrument called a *pH meter*, which typically determines the acidity or alkalinity of a solution/liquid. The major components of a pH meter are a measuring probe (recognition element), that is usually a glass electrode, and an electronic meter (transducer) to take and display the measurement, as illustrated in Figure 4.8.

pH meters work on the basis of the generation of small voltages by two electrodes (comprising a probe), which are passed on to the meter and converted into pH units for display. Historically, these electrodes were wholly separate units, the first being a pH-dependent glass electrode (sensitive to H^+ ions), and the second a pH-independent reference electrode (comprising mercury(I) chloride). When used together, these two electrodes created a potential difference measured as a voltage which, importantly, depends on ambient temperature. This means that adjustments needed to be made to the pH meter to reflect the temperature, usually by means of a control knob. Modern instruments use a combined pH probe which incorporates both the pH measuring electrode and reference electrode.

Combined measuring probe: Has a thin-walled glass bulb at its tip which detects H^+ ions in solution, and produces a small voltage (around 0.06 V per pH unit), and as such provides a signal related to the concentration of H^+ ions which can be converted into a measurement of pH. For simplicity, the pH probe can be

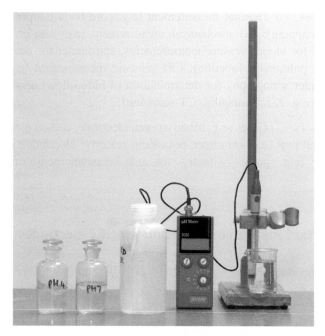

Figure 4.8   Photograph of a pH meter.

considered as a galvanic cell comprising a tube within a tube. The innermost tube is the reference electrode, which has a constant composition (i.e. unchanging solution), whereas the outermost tube contains an ion-rich medium, which is allowed to mix with the outside environment (serving as salt bridge). It is the mixture and transfer of ions from the glass bulb (ion-selective electrode) that creates a change in energy, and it is this change that is measured by the pH meter. As there is mixing, the outer tube must be periodically replenished with KCl; the pH probe should be handled with care to avoid damage, and particular attention must be given to keeping the glass bulb hydrated (e.g. by immersing in 0.1 M HCl or 0.1 M H_2SO_4).

Electronic meter: Essentially a voltmeter which rather than displaying volts gives measures of pH. This is possible because the small voltages produced by the H^+ ion concentration directly relate to pH. The meter contains an inverting amplifier that converts the small voltages produced by the probe into pH units which are then offset by 7 V to give a pH reading. This offset relates to the fact that at neutral pH the probe's voltage output is zero and can be adjusted during calibration.

Using a pH meter: Before taking pH readings it is important that the user checks the components of the system, paying particular attention to the *measuring probe* (e.g. level of KCl and integrity of glass bulb) and *temperature control* (which should match the temperature of the solution). The measuring probe, which is maintained in a storage solution, is removed, rinsed well with distilled water and gently blotted dry before use. As a first step the instrument needs to be calibrated, and this is achieved by immersing the electrode in at least two different standards. These standards are popularly commercially available buffer solutions which are of neutral (pH = 7.01), acidic (pH = 4.01) or alkaline pH (pH = 10.01), which are valid at 25 °C. Once the gain and offset parameters of the meter have been adjusted to the pH of one standard (i.e. meter pH reading matches pH reading of standard) by means of calibration knob(s), the probe is then removed, rinsed and transferred into the second standard for a second calibration step. This process is repeated until accurate readings are obtained with both standards and the meter is then ready to use with the unknown test solution. In cases where the pH of the unknown test solution can be roughly estimated before reading, typically a pH standard is employed which is within 2 pH units of the unknown solution.

Ion-selective and redox electrodes

The measurement of pH, as described above, utilizes an ion-selective electrode; there are other types of ion-selective electrode which can be attached to meters for display/output. As such, and consistent with the workings of a pH meter, an ion-selective electrode is a transducer which converts activity of a specific ion in solution into an electrical potential which is measured by a voltmeter. Ion-specific electrodes typically use ion-specific membranes (recognition element) and a reference electrode, and are used for measures of ionic concentration in real time. Perhaps the most important clinical uses are for determination of ions in body fluids, including sodium (Na^+), potassium (K^+), magnesium (Mg^{2+}) and calcium (Ca^{2+}), and these measures give important diagnostic information. Redox electrodes are primarily used for measuring redox potentials generated by electron transfer reactions in a specific redox system, but may also be used in other electrochemical studies. The electrode is made from high stability electron-conductive material, and these may be either inert metal electrodes (e.g. platinum or gold electrodes) or metal electrodes (e.g. silver/silver chloride electrode).

4.5 Specific applications of biosensors in life and health sciences

Biosensors are particular types of chemical sensors that specifically utilize a biological recognition element, converting biochemical signals into quantifiable electrical signals. In biosensors, the recognition element is typically an immobilized biological substance (analyte) in intimate contact with a physicochemical transducer. Biosensors comprise a sensitive biological recognition element–either biocatalytic reaction (e.g. enzyme-based) or binding process (e.g. affinity-based), which is typically bioengineered, coupled with a physicochemical transducer/detector (e.g. electrochemical or optical device). Some examples of important applications of biosensors are outlined below.

Blood glucose meter

Insulin is an important metabolic hormone that lies at the heart of whole body metabolism, with a particularly important role in regulating circulating blood glucose levels. Given this, deficiencies in insulin production, secretion or action on target tissues results in improper glucose utilization and high levels of blood glucose (hyperglycaemia), and impaired glucose tolerance and diabetes. The most widespread use of biosensor technologies in medical practice is the glucose meter routinely used by diabetic patients to monitor their blood glucose levels (self-monitoring). The generic term *glucose meter* encompasses a large number of devices, which may be bench-top, or more conveniently hand-held, utilizing different chemical reactions to provide readings of glucose in very small volumes of blood.

Most devices rely on the use of disposable one-use-only 'test strips', and a droplet of blood from a needle-prick is drawn, for example, from the thumb, onto the strip by capillary action. The strip represents the recognition element, and the blood sample undergoes an enzymatic chemical reaction followed by another reaction. After the blood is added to the test strip it is inserted into the hand-held device (transducer) and sufficient time allowed for the reactions to occur, most often automatically determined by the device itself.

An example of a test strip and hand-held recording device is shown in Figure 4.9. Blood is added to one end of the test strip and drawn between two electrodes (measuring electrode and reference electrode) and the strip inserted into the device.

There is first a chemical reaction, usually involving oxidation of glucose and reduction of an enzyme, popularly glucose oxidase. In this reaction, glucose oxidase catalyses the oxidation of beta-D-glucose to D-glucono-1,5-lactone, which is subsequently hydrolysed to D-gluconic acid. This reaction is then followed by another, most commonly an electrochemical reaction, and as a result of these two reactions a small electrical current is produced which directly relates to the amount of glucose in the sample. This current is processed by the transducer (hand-held device), recorded in a short period (usually 30 s) and displayed in appropriate units of blood glucose–millimolar in the United Kingdom/Europe and milligrams per

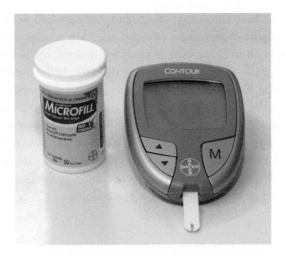

Figure 4.9 **Example of a typical blood glucose meter and test strip.**

decilitre in the United States. Importantly, each device will come with reference strips used to assess the performance and reliability of the workings of the meter.

Electronic nose

Another emerging biosensor technology is the electronic nose, a generic term used to describe sensor devices that, as the name suggests, detect and recognize particular odours and flavours. While the stages of recognition are essentially similar to the human nose (human olfaction), importantly these devices, despite using sensor arrays and pattern recognition systems, are unable to make the same subjective determinations. Despite inherent limitations, these devices have found numerous industrial applications and are useful alternatives to human sensory analysis or other bioanalytical tools such as gas chromatography (GC, GC/MS). Essentially, the electronic nose functions with recognition element(s) and transducer/signal processor. Most electronic noses use sensor arrays that interact with volatile/gaseous compounds, whereby adsorption of the compound onto the surface of the sensor causes a physical change (conductivity, resistance and frequency) and the signal is digitized and processed by computer, which performs a global fingerprint analysis allowing characterization of complex mixtures and indicating what substances are present. As such, electronic noses are previously 'trained' with known samples, building the reference database for qualitative and quantitative analyses.

While electronic nose technology has been largely applied in the food industry to characterize odours of beverages or olive oil, such sensors are not limited in their application in other fields such as medicine. Indeed, it has long been known that smell was an early means of diagnosing diseases, being used by the Ancient Greeks and Chinese. Ongoing research has been directed towards how electronic noses could be applied, and one good example is detection and discrimination of volatile compounds associated with bacterial infections in humans and animals, potentially allowing diagnosis on the basis of a simple analysis of breath. As such, the ultimate goal is to make inexpensive portable point-of-care devices that can be used as effective bioanalytical tools, for example in the diagnosis of diverse lung infections and in monitoring disease epidemiology.

Key Points

- An ion is an atom (or group of atoms) that has lost or gained one or more electrons through oxidation or reduction.

- Oxidation is the loss of one or more electrons (i.e. oxidation is loss; OIL), reduction is gain of one or more electrons (i.e. reduction is gain; RIG), that is, OIL RIG.

- An oxidizing agent removes electrons from another substance (i.e. causes oxidation), thereby gaining electrons (i.e. is an electron acceptor), while a reducing agent gains electrons from another substance (i.e. causes reduction), thereby losing electrons (i.e. is an electron donor).

- An acid is a proton donor, dissociating in water to form hydrogen ions or protons, while a base is a proton acceptor, dissociating in water to form hydroxide ions.

- pH scale is a numerical representation of the acidity or alkalinity of a solution.

- pH is measured using a pH meter, which primarily consists of a measuring probe (recognition element) and an electronic meter (transducer).

- Buffer solutions comprise a weak acid or base and its salt (conjugate acid or base), and provide a means of resisting or modulating changes in pH of solutions, in order to keep the environment constant to allow optimal conditions for biochemical reactions.

- A physical sensor measures physical quantities including electricity, length, temperature and weight.

- A chemical sensor selectively measures specific chemical substance(s) in a sample in a qualitative or quantitative manner.

- A biosensor is a particular type of chemical sensor, which specifically utilizes a biological recognition element, converting biochemical signals into quantifiable electrical signals.

5 Applications of spectroscopy

The various forms of spectroscopy represent fundamental tools for the bioanalytical chemist. Collectively these important technologies rely mainly on the different light absorbing and emitting properties of biomolecules, usually in solution. However, spectroscopy is much broader than this, representing the study of matter through measures of light, sound, or particles that are absorbed, emitted or scattered. As the most popular use of spectroscopy is to study the interaction of light and matter, this will be the primary focus of this chapter. The history of spectroscopy dates back to the 1600s when Sir Isaac Newton first reported the dispersion of white light into a spectrum of colours using a basic triangular glass prism. In the late 1800s Bunsen and Kirchhoff added a few lenses and a 'slit' to generate the now classical prism spectroscope. Since then, spectroscopes have evolved from the use of a slit to incorporate a diffraction grating, allowing dispersion of light, and a light sensor (photodetector) facilitating more sophisticated analyses. While spectroscopes are still used for particular applications (e.g. in astronomy), they have now largely been superseded by spectrometers that allow more diverse measures and applications. The following sections provide an overview of the major types and modern applications of spectroscopy.

Learning Objectives

- To understand the basic principles underlying spectroscopic techniques.
- To distinguish between various major types of spectroscopy.

Understanding Bioanalytical Chemistry: Principles and applications   Victor A. Gault and Neville H. McClenaghan
© 2009 John Wiley & Sons, Ltd

- To describe the principles, methodology, instrumentation and applications of ultraviolet (UV)/visible spectroscopy.

- To discuss the principles, methodology, instrumentation and applications of infrared (IR) spectroscopy.

- To summarize the principles and applications of fluorescence spectrofluorimetry.

5.1   An introduction to spectroscopic techniques

In order to understand the principles and applications of spectroscopy in bioanalysis it is important to appreciate some fundamental aspects of waves and electromagnetic radiation. At the heart of spectroscopy, spectrometry and spectrophotometry is the interaction of waves of energy (electromagnetic radiation and non-electromagnetic radiation) and matter in a sample to be analysed. Matter will interact with these waves of energy in different ways, and it is this diversity that is important with respect to identification, analysis and characterization of biomolecules in a sample.

Basic spectroscopy

The earliest form of spectroscopy (used in its loosest sense) was reported by Sir Isaac Newton, who when sending a single beam of white light source through a glass prism noted that this light was scattered/refracted to generate the classic visible rainbow pattern of seven colours (red, orange, yellow, green, blue, indigo and violet). Each of these component colours represents a defined wavelength of light and, in essence, a spectroscope simply disperses light by wavelength but cannot inherently quantify or distinguish the brightness/intensity of each emitted colour. To address this deficiency, so-called spectrometers were developed, which had the added value of being able to provide measures of the intensity of a given wavelength of emitted light. Modern forms of spectroscopy can allow the measurement of more than one wavelength (e.g. dual or scanning spectrometer) or indeed utilize, measure and compare more than one beam of light (so-called spectrophotometers).

Waves and electromagnetic radiation

Characteristically, energy is absorbed, emitted or scattered in the form of waves. These energy waves have certain fixed features that include wavelength and frequency. These values are inversely proportional to each other and can be defined by the following equation:

$$c = \lambda v$$

where c is the speed of light ($3 \times 10^8 \, \mathrm{m \, s^{-1}}$); λ is wavelength of the radiation (pronounced Lambda); v is frequency of the electromagnetic radiation (Hz).

As the speed of light is a constant, from this equation you can see that the shorter the wavelength (i.e. lower the value of λ), the greater the frequency of electromagnetic radiation and the higher its energy.

The wavelengths of different colours of light are represented in the visible range of the electromagnetic spectrum. However, as indicated in Figure 5.1, the electromagnetic spectrum covers a very wide range of different wavelengths/frequencies which are important for laboratory-based spectroscopy, representing numerous types of energy waves ranging from X-rays to radiofrequency.

In spectroscopy it is useful to consider the propagation of electromagnetic radiation in a quantitative manner. Light is transmitted as 'discrete packets' or as a 'stream of particles' of energy called photons. These photons have a specific energy and for spectroscopy are quantized and described by the following equation:

$$E = hv$$

where E is energy; h is Planck's constant ($6.626 \times 10^{-34} \, \mathrm{J \, s^{-1}}$); v is frequency of the electromagnetic radiation.

In spectroscopy, this quantized energy in the form of photons is applied to biomolecules in a sample and energy is exchanged, and the change in energy level of the biomolecule can be measured.

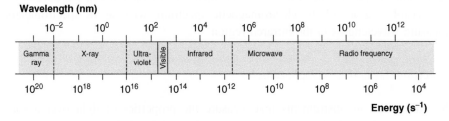

Figure 5.1 **Diagrammatic representation of the electromagnetic spectrum.**

Interaction between radiation and matter

Spectroscopy allows study of the exchange of energy between electromagnetic radiation and biomolecules to which it is applied, relying on the production and recording of electromagnetic radiation. The transmission of energy, in the form of electromagnetic waves characterized by their frequency and wavelength, can be recorded and analysed. An electromagnetic spectrum arises when atoms in a sample absorb and emit electromagnetic radiation. When atoms absorb energy as electromagnetic radiation they move from an unexcited 'ground state' to an 'excited state' (sometimes referred to as the *first excited state*), whilst atoms which emit energy as electromagnetic radiation move from an 'excited state' to a lower excited state, and when they give up all their energy they reach the 'ground state'. This exchange of energy between lower- and higher-energy states is of fundamental importance to spectroscopic analysis, and is described by the following equation:

$$\Delta E = h\nu$$

where ΔE is the difference in energy (e.g. change in energy from one state to another, i.e. $E_2 - E_1$); h is Planck's constant ($6.626 \times 10^{-34}\,\mathrm{J\,s^{-1}}$); ν is frequency of the electromagnetic radiation.

From this equation it can be seen that high energy jumps require electrons to absorb light of a higher frequency, that is, the greater the frequency, the greater the energy. In practical terms, spectroscopy measures the discrete amount of energy exchanged between electromagnetic radiation and biomolecules. Given this, data arising from spectroscopy gives important structural information regarding the biomolecules present in a sample. However, it is essential to note that laboratory spectrometers/spectrophotometers do not directly measure energy or frequency, instead recording wavelength. As such, it is important to appreciate the relationship between frequency and wavelength ($\lambda = c/\nu$), where the higher the frequency the lower the wavelength. It is the difference in the wavelength being measured that helps define the most commonly encountered spectrometers, that is, UV/visible spectrometers (sometimes called UV/vis) measure wavelengths in the ultraviolet and visible regions of the electromagnetic spectrum, whereas IR spectrometers measure wavelengths in the infrared region.

Spectrometer and spectrophotometer

Spectrometers are instruments that measure the properties of light over a specific region of the electromagnetic spectrum, determining light intensity and other parameters such as its polarization state. Inherently, a spectrometer utilizes a single beam of light to produce spectral lines from which it determines wavelength and intensity.

The main type of spectrometer used in routine bioanalysis is an absorption spectrometer which, as the name suggests, measures absorption of light, that is, the energy from the photons. All absorption spectrometers share the following common features and components: (i) a radiation source; (ii) a monochromator (diffraction grating); (iii) sample container (sometimes called a *cell* or *cuvette*); (iv) frequency analyzer; (v) detector; and (vi) recorder. Absorption spectrometers can vary in their construction depending on, for example, the radiation source and form of electromagnetic radiation and the nature of the biomolecules being analysed.

As the name implies, a spectrophotometer is a spectrometer which measures intensity as a function of wavelength (or colour) of light (absolute light intensity). Two major types of spectrophotometer are the so-called single-beamed and dual-beamed instruments, illustrated in Figure 5.2. A dual-beamed spectrophotometer is an instrument where two separate radiation sources are used to generate two individual beams, which interact with a sample and a reference and produce two beams of emitted light; the spectrophotometer can then compare the intensities of these two beams of light (so-called relative absorbance).

There are a number of ways to classify spectrophotometers, but the major distinctions relate to (i) wavelength they analyse; (ii) methods of measurement; (iii) means of data acquisition (i.e. how they acquire a spectrum) and (iv) the ability to differentiate between intensities (i.e. intensity variation they can measure). Spectrophotometers generate absorbance measurements which follow the Beer–Lambert law described later.

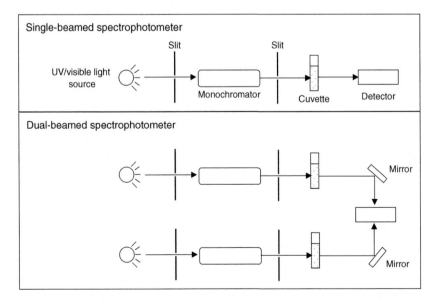

Figure 5.2   Schematic of typical single- and dual-beamed spectrophotometers.

The choice of spectroscopic technique (spectrometer or spectrophotometer) to be employed depends on the physical aspect to be measured and whether or not continuous recording over a range of wavelengths is required. Importantly, spectrometers differ in their resolving power, that is, their ability to distinguish between two particular frequencies or wavelengths of absorption. Spectroscopic techniques can be generally categorized as absorption spectroscopy, emission spectroscopy, and scattering spectroscopy. However, it is important to note that electromagnetic radiation can interact with matter in other ways and other instruments including nuclear magnetic resonance (NMR) and electron spin resonance (ESR) described in Chapter 11 can record this.

5.2 Major types of spectroscopy

Absorption spectroscopy: Records the electromagnetic radiation (energy) as atoms absorb energy and thus are promoted to a higher energy level (i.e. 'excited'). Each biomolecule will have a specific absorption spectrum, which can act as a fingerprint, and the more of the biomolecule present the higher the absorption, thus allowing analytical identification and quantification. Examples of absorption spectroscopy include atomic absorption (see Chapter 3), UV/ visible (see later), infrared (see later), and Mossbauer (allowing determination of absorption of gamma-rays).

Emission spectroscopy: Records the electromagnetic radiation (energy) emitted as atoms undergo transition to a lower energy level. In this case biomolecules give off low levels of energy, commonly in the form of fluorescent light representing discrete photon emissions. Primary examples of emission spectroscopy include fluorescence (see later), atomic or flame emission (where energy is provided to the sample through use of a flame), X-ray fluorescence (where energy is provided to the sample through use of X-rays and fluorescent X-rays are emitted) and stellar (used in astronomy where emitted light produces stellar spectra sharing features of the Sun's optical emission spectrum).

Scattering and other forms of spectroscopy: Rely on the fact that electromagnetic radiation has other interactions with matter beyond that of simple absorption and emission. These interactions generate other measurable quantities such as scattering of polarized light (e.g. circular dichroism), and changes of spectral features of chemical bonds (e.g. Raman spectroscopy).

These two forms of spectroscopy are commonly categorized under scattering spectroscopy which, as the name suggests, measures the amount of light that a sample scatters at defined wavelengths/incident angles/polarization angles. This type of spectroscopy is different from absorption and emission spectroscopy as the scattering process is much quicker than the processes of either absorption or emission. Notably, however, there are other important forms of spectroscopy which do not rely on absorption, emission or scattering. This includes resonance spectroscopy, which records changes in magnetic states of atomic nuclei or electron spin (e.g. NMR, ESR; see Chapter 11).

As mentioned above, spectroscopy is an umbrella term for a range of techniques that allow identification, quantification and determination of molecular structure of biomolecules. This technology also allows the bioanalytical chemist to follow key reaction pathways including, for example, the rate of an enzyme-catalysed reaction.

5.3 Principles and applications of ultraviolet/visible spectrophotometry

Background and principles: UV spectroscopic analysis was one of the first applications of spectroscopy in an analytical context. While a powerful technique of yesteryear, this method is now rather restricted in its use in a modern bioanalytical laboratory. However, UV/visible spectroscopy is still used in specific biochemical analyses, such as dyestuffs in forensic applications. Despite these limitations, a bench spectrophotometer that measures UV/visible absorbance is commonly found in both teaching and research laboratories. In order to better understand the underlying principles it is necessary to consider the *excitation* of electrons, electronic transition and how this is related to energy and *absorbance*.

A biomolecule can be excited by photons, which will cause electronic transition between energy states. In the case of excitement, most often molecular electronic transitions occur when valence electrons in a biomolecule are excited from one energy level (e.g. ground state) to a higher energy level. Absorption will only occur when the energy of the photon corresponds exactly with the energy required to reach the next transition state, and the relationship between energy involved in electronic transition and frequency of radiation is given by the equation $\Delta E = h\nu$, considered earlier. From the energy acquired through transition, a number of important features can be determined, including structural information and molecular and other properties including colour. The structural parts of biomolecules, which permit absorption in the UV/visible region, are referred to as *chromophores*, to which a group of atoms (auxochromes) may be attached.

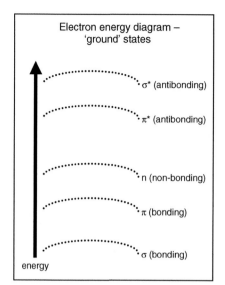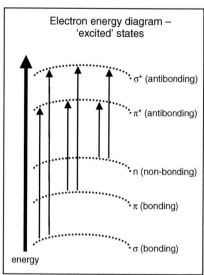

Figure 5.3   **Diagram illustrating possible promotion of electrons on absorbance of light.**

Auxochromes (e.g. –OH–NH$_2$) modify the ability of a chromophore to absorb light by virtue of the interaction of their lone pair of electrons with the pi-system (π) of the chromophore (the so-called mesomeric effect), and may intensify the absorption. Given this, a biomolecule with its characteristic electronic structure and bonding pattern will absorb light at a defined frequency. The energetics of bonding are very complex, but essentially all bonds and electrons involved in the process of bonding have particular features which can be assessed by UV/visible spectroscopy, providing the transitions are in the UV or visible range of the electromagnetic spectrum. Electrons exist in various different orbitals, circulating around the atom's nucleus. The energy states of the orbitals are illustrated in Figure 5.3.

As shown, when excited by energy from light, an electron can jump from a so-called bonding orbital to a so-called antibonding orbital. These are often referred to as the *highest occupied molecular orbital* (HOMO) and the *lowest unoccupied molecular orbital* (LUMO). When considering sigma (σ) bonds, electrons in the HOMO can be excited to the LUMO and this process is described as a bonding to antibonding transition, denoted as $\sigma \rightarrow \sigma^*$. In pi ($\pi$) bonds, electrons can also be promoted from a so-called π-bonding orbital to a π^*-antibonding orbital, written as a $\pi \rightarrow \pi^*$ transition. Free electron pairs (non-bonding orbital) can also undergo transitions and these are described as either $n \rightarrow \sigma^*$ or $n \rightarrow \pi^*$ transitions. Electrons in π bonds and lone pairs (n) require less energy than σ bonds for excitation, and as such absorb at longer wavelengths. Measurements in UV/visible spectroscopy are carried out at wavelengths of around 200–300 nm (UV range) and around 330–700 nm (visible range) of the electromagnetic spectrum.

Fundamentals of spectrophotometer readings: In order to best understand the use of UV/visible spectrophotometry and a spectrophotometer instrument it is important to first understand the data that can be derived from the use of this technique/equipment. The fundamental outcome from the use of a spectrophotometer is a measure of transmittance or absorbance. As the names suggest, transmittance (T) is the amount of light that is *transmitted through* the sample solution whereas absorbance (A) is a measure of light *absorbed by* the sample solution. Modern instruments can provide readings of both *transmittance* and *absorbance* (or its reciprocal $1/A$); normally the primary reading for most bioanalytical applications is absorbance. However, it is important to appreciate that transmittance and absorbance are related by fundamental equations.

As the buffer in which the sample is dissolved may itself absorb light at the same wavelength as the sample to be analysed, it is necessary to compare the light intensity readings from both the sample solution and a reference blank (i.e. same solution but without sample). Transmittance is a measure of the ratio between the light intensity reading of the sample (I) and the light intensity reading of the reference blank (I_0). These measures inherently assume that the reference blank does not contain any sample, and thus the transmittance of the reference blank is 100% T. However, transmittance is often expressed as a percentage (percentage transmittance), and logically %T is lower than for the reference blank (100% T):

$$T = I/I_0$$

$$\%T = (I/I_0) \times 100$$

For the majority of bioanalytical applications, measures of absorbance will be made at defined wavelengths, and the absorbance reading is sometimes referred to as the *optical density* (OD) reading. The absorbance is calculated from the transmittance measure using the logarithmic relationship described by the following equation:

$$A = \log(I_0/I) = -\log T$$

A popular use of a spectrophotometer is to measure concentration using the relationship between concentration and absorbance described by the *Beer–Lambert law*, which dictates that the absorbance of a light-absorbing substance is proportional to its concentration in solution:

$$A = \varepsilon c l$$

and

$$A = \log(I_0/I) = \varepsilon c l$$

where A is absorbance at a particular wavelength (sometimes denoted A_λ); ε is the extinction coefficient of the substance (units $M^{-1}\,cm^{-1}$); c is concentration of the substance in the sample solution (units M); and l is path length through sample solution (units cm; width of cuvette – typically 1 cm).

Importantly the Beer–Lambert law equation can be rearranged to give different measures, providing a useful method for quantitative and qualitative analyses, so:

$$A = \varepsilon c l \quad \text{and} \quad A/\varepsilon l = c \quad \text{and} \quad A/cl = \varepsilon$$

Notably, spectrophotometers will only give meaningful absorbance readings within a defined range, that is, the absorbance readings become unreliable and inaccurate if they are too high or too low. To avoid this simple error, absorbance readings should really lie within the range 0.020–1.5 for an instrument that measures between 0 and 2, but obviously the upper detection limit will depend on the instrument used. If the reading is too low then the sample is too dilute and often this can be corrected simply by adding more of the sample to the solution (i.e. higher concentration). Conversely, if the reading is too high then the sample is too concentrated and usually this requires a simple dilution of the sample in solution (i.e. lower concentration). Also, in most cases there will be an optimum wavelength for measuring absorbance of a particular substance in solution, that is defined as A_{max}. Another term you may come across is ε_{max} that provides additional structural information.

Basic equipment and components: As noted earlier, generally spectrophotometers rely on the use of a dual beam (or double beam), and are the most commonly encountered spectroscopic instruments in bioanalytical laboratories. The simplest spectrophotometers have a single beam (e.g. Spectronic 20), where the entire light beam passes through the sample, so for the relative absorbance two separate readings are required. The first is the absorbance of the reference, which is usually the solution (e.g. buffer or water) without dissolved sample, and the second absorbance reading is for sample in the solution. This form of spectrophotometer is commonly seen in teaching laboratories. In contrast, a dual-beam UV/visible spectrophotometer splits the UV and visible beam into two identical beams, one of which passes through the sample in a solution (buffer or water) while the other passes through a reference (normally the solution without sample – a control). Each beam is then focused onto a detector and the ratio of the intensities of the two beams is measured, where the difference in the intensities is converted into an absorbance reading. As noted earlier, the intensity reading of the reference is given by I_0, and the reading from the sample is I, where ratio I/I_0 is called the *transmittance*. The absorbance value is then determined from this measure of transmittance, as noted earlier. The light source for visible wavelengths is an incandescent bulb (usually tungsten-halogen) and UV light is usually emitted from a deuterium lamp. Sample is normally contained in a transparent rectangular cell (plastic/glass or preferably quartz cuvette) of defined width, through which the light passes (i.e. path length; typically 1 cm). Light separation (into different wavelengths) is achieved using a *diffraction grating* or *monochromator*, and the detector is usually a *photodiode* or *charge-coupled device* (CCD), each of which respond to light with generation of an electrical current. Typically, photodiodes are used with monochromators, filtering light so that only light of single wavelength reaches the detector, whereas CCDs are used with diffraction gratings, where the detector can collect light of different wavelengths.

Methodology: UV/visible spectrophotometry can be used to determine a number of variables. As outlined in the equations above, various measures can be taken to ultimately derive various parameters; primarily absorbance (A), concentration (c) and extinction coefficient (ε). The following example illustrates how to make basic readings using a bench spectrophotometer. The spectrophotometer instrument must be switched on for around 15–30 min before readings are made, to allow the bulbs to warm up. Wavelength (in most instances λ_{max}) is then selected using the electronic control panel, and the spectrophotometer adjusted to transmittance mode before setting percent transmittance to zero using the controls. A clean, dry quartz cuvette is filled to around 4/5 full with reference blank (i.e. solution without sample) and carefully inserted into the holder inside the instrument. It is vital that the cuvette is orientated in the correct position so the beam can effectively pass through the cuvette (often this simply means lining up the mark on the cuvette with the mark on the holder). The spectrophotometer is then adjusted to absorbance mode and the absorbance reading set to read zero using the controls. After this adjustment, the reference blank is replaced with the cuvette containing the sample in solution. An absorbance reading for the sample in solution can then be recorded. For convenience (and only if the operator is convinced that the instrument is reliable) a number of absorbance readings at the same wavelength can be recorded in quick succession without using the reference blank each time. However, if absorbance readings are to be taken at different wavelengths it is imperative that the reference blank be used for each independent wavelength.

Important notes: If a sample is prepared in a denaturant which contains a light-absorbing compound, it is necessary to add the same concentration of this denaturant to the reference blank to avoid inaccuracies in absorbance readings. Also, when determining the concentration of substance in a sample in solution it is crucial to measure absorbance at multiple wavelengths, including one where absorbance is anticipated to be zero. This helps determine the presence of possible contaminants. For example, to determine protein concentration, absorbance readings should ideally be taken at 320 nm (absorbance close to zero), 280 nm (near A_{max}) and 260 nm (typically detecting nucleic acid contaminants). Notably, these same wavelengths are used for absorbance readings of DNA and RNA. Other considerations include ensuring there are no air bubbles or particulates in the sample solution and that the sample solution is thoroughly mixed before transfer to the cuvette. While single readings attained by using the method above are useful, more advanced spectrophotometers allow scanning and generation of an absorbance spectrum requiring more detailed analysis and data interpretation.

Interpretation of absorbance spectra: More complex analyses are available from spectrophotometers that allow scanning and generation of an absorbance spectrum. A UV/visible absorbance spectrum is basically a graph of absorbance versus wavelength; however, it is important to remember that plots can also be made of other parameters (e.g. extinction coefficient versus wavelength). The point in the graph where the maximum absorbance is recorded is called λ_{max}, the wavelength reading taken at the top of an absorbance spectrum curve. However, another important measure is the 'extent of absorbance' which is the absorbance reading at a given concentration of substance at a given wavelength. So, if the concentration of substance is higher, the λ_{max} reading will be the same, but the absorbance reading will be higher, thus giving a higher extent of absorbance (see Figure 5.4).

Each compound will have particular bonding characteristics, which will generate different values of λ_{max}. As noted earlier, bonding depends on the behaviour of electrons and electronic transitions. These electronic transitions will be observed at different wavelengths on the absorption spectrum and reveal important features of the compounds and chromophores being studied. Values of λ_{max} and absorbance can be taken from the absorbance spectrum, and from this the maximum extinction coefficient (ε_{max}) can be determined using the Beer–Lambert law equation. Table 5.1 gives information on electronic absorption characteristics of typical chromophores in non-aromatic compounds.

Unlike non-aromatic compounds, aromatic substances display discrete bands associated with electronic transitions. These bands are defined according to the

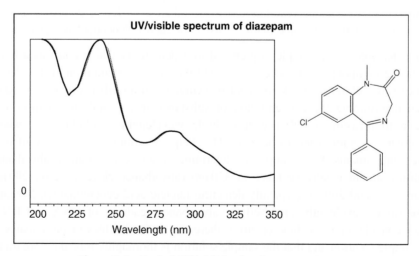

Figure 5.4   **Typical UV/visible absorbance spectrum.**

Table 5.1 Absorption characteristics of some
typical non-aromatic chromophores

Example	Transition	λ_{max} (nm)
Ethylene	$\pi \rightarrow \pi^*$	165
Acetone	$\pi \rightarrow \pi^*$	188
	$n \rightarrow \pi^*$	279
Azomethane	$n \rightarrow \pi^*$	347
Nitromethane	$n \rightarrow \pi^*$	278

Table 5.2 Absorption characteristics of some
typical aromatic chromophores

Example	Transition	λ_{max} (nm)
Benzene	$\pi \rightarrow \pi^*$	184
		200
		255
Acetophenone	$\pi \rightarrow \pi^*$	199
		246
		279
		320
Nitrobenzene	$\pi \rightarrow \pi^*$	252
		280
		330
Naphthalene	$\pi \rightarrow \pi^*$	221
		286
		312

system devised by Burawoy in the 1930s, and are called *R-band* (radical-like), *K-band* (conjugated), *B-band* (benzoic acid), and *E-band* (ethylenic). For example, the common organic aromatic compound, benzene, has three bands representing three aromatic $\pi \rightarrow \pi^*$ transitions, namely two E-bands (at ~180 and ~200 nm) and one B-band (at ~255 nm). Table 5.2 gives information on electronic absorption characteristics of typical chromophores in aromatic compounds.

These tables provide simple examples of commonly encountered compounds, but of course there are a huge number of complex biomolecules that can be characterized by UV/visible spectrophotometry. Such characterization is primarily based on the presence of distinct functional groups.

Applications: As noted earlier, UV/visible spectrophotometers are commonly encountered laboratory tools with a number of fundamental applications, several

of which are outlined as follows: (i) *Structural analysis:* UV/visible spectrophotometry can be used to identify chromophores based on measures of λ_{max} and
ε_{max}, which identifies structural features of biomolecules. When used in conjunction with other spectroscopic methods (e.g. IR), UV/visible spectrophotometry
can provide useful information on determining the exact molecular structure of
biomolecules such as steroid hormones. For example, testosterone contains adjacent C=C and C=O double bonds, generating a characteristic λ_{max} of ~240 nm.
(ii) *Quantitative analysis:* by taking a number of absorbance readings over a
concentration range (serial dilution of known sample) it is possible to construct a
standard curve (of absorbance versus concentration), from which an unknown concentration can be determined from an absorbance reading. In addition, UV/visible
spectrophotometers can be used to provide kinetic information, that is, determination of the absorbance (at a set wavelength) versus time, which allows measures of
the loss or gain of a biomolecule in a sample solution. Given this, the spectrophotometer not only provides information on the product(s) of a chemical reaction
but also the rate of the reaction. (iii) *Pharmaceutical analysis:* as UV/visible spectrophotometry can give definite measures of ε_{max} under set conditions (i.e. specific
concentration, path length and solvent used), this method enables quantification
of a particular drug in a tablet/capsule formulation. Also, given this, it is possible
to use this approach for quality assurance, which is particularly important in the

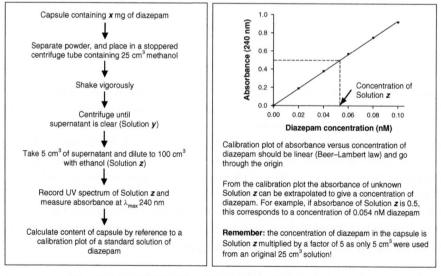

Figure 5.5   **Application of UV/visible spectrophotometry in drug analysis and quantification.**

pharmaceutical industry to control drug dosing and contamination. For a simple example of this in practice see Figure 5.5.

5.4 Principles and applications of infrared spectroscopy

Background and principles: Infrared radiation describes electromagnetic radiation at frequencies and energies lower than those associated with visible light. The IR region of the electromagnetic spectrum is subdivided into three parts, namely the near-, mid- and far-infrared, given as wavenumbers (that is wavelengths per cm, units cm^{-1}), describing how they relate to the visible spectrum. Far-infrared ($400-10\,cm^{-1}$) lies adjacent to the microwave region of the electromagnetic spectrum, is low energy and can be used to study rotational spectroscopy. Mid-infrared ($4000-400\,cm^{-1}$) is used to study fundamental molecular vibrations described in detail below, and near-infrared ($14\,000-4000\,cm^{-1}$), of high energy, can be used to generate overtone or harmonic vibrations. As IR spectroscopy depends on compounds absorbing energy, the major principles are essentially the same as those of UV/visible spectroscopy outlined earlier. However, importantly, a fundamental difference lies in the fact that absorbed energy causes vibrational excitation of bonds in a compound, as opposed to the electronic transitions resulting from absorbed UV/visible light. IR spectroscopy relies on the specific frequencies at which chemical bonds vibrate or rotate corresponding to the energy levels applied to a compound. Bonds can be excited by IR radiation to cause bond 'stretching' (higher energy) or bond 'bending' (lower energy) vibrations as illustrated in Figure 5.6.

Furthermore, stretching or bending of bonds can be further classified into various vibrational modes. In the case of stretching, the modes can either be 'symmetrical' or 'asymmetrical' ('antisymmetrical') and modes of bending include 'scissoring', 'rocking', 'twisting' and 'wagging'. Even moderately sized compounds contain groups such as $-CH_2$ or $-NH_2$, which will possess these different vibrational modes. Furthermore, when considering molecular vibrations, even simple molecules have many bonds, each of which can stretch and bend independently, resulting in complex spectra, and as such certain simplifications are necessary. In addition, for IR, as asymmetry gives a much stronger signal it is generally considered a requirement, given that symmetrical vibrations only give rise to weak signals. Also, for molecules to absorb IR there must be a change in dipole moment as described later.

To better understand the nature and features of these vibrations, bonds can be considered as springs. Given this analogy, the behaviour of these molecular springs approximately follows Hooke's law of elasticity. In physics, Hooke's law relates the 'strain' on a body (spring) to the force (load or mass) causing the 'strain'. In essence, molecular bonds follow this linear relationship, where the

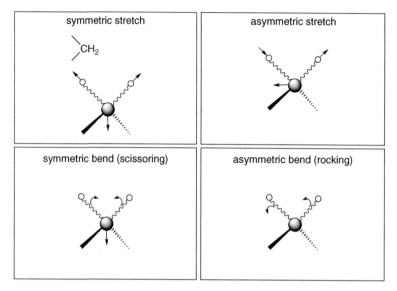

Figure 5.6   Illustration of vibrational modes in IR spectroscopy. (Adapted from Silverstein *et al.* Spectrometric Identification of Organic Compounds, 7th edn, 2005.)

extension of the bond (or spring) is directly proportional to the force applied to the bond. Derivation of the following equation, which correlates vibrational frequency with bond strength and atomic mass, from the Hooke's law equation involves some serious mathematics which we would rather not expose the reader to in this textbook:

$$\nu = \left(\frac{\pi c}{2}\right)\sqrt{\left(\frac{k}{\mu}\right)}$$

where ν is vibrational frequency (Hz); c is the speed of light $(3 \times 10^8 \, \text{m s}^{-1})$; k is force constant of the bond (bond strength, units Newtons per metre or $N \, m^{-1}$); μ is reduced mass (kg).

Reduced mass relates to the relationship of the masses (atoms) at either end of the molecular spring, termed m_1 and m_2, where reduced mass is defined as $(m_1 \times m_2)/(m_1 + m_2)$. As an example, in a C–H bond the masses in question are $C = m_1$ and $H = m_2$.

Considering the above equation and remembering that energy also relates to frequency $(E = h\nu)$, a number of important points should be noted:

- Energy is directly proportional to bond strength.

- Bond stretching requires more energy than bond bending, and as such bond stretching absorbs shorter wavelength, higher frequency radiation than bond bending.

- When exciting vibration in bonds, single bonds require (and absorb) less energy than double bonds, which require (and absorb) less energy than triple bonds.

- Energy is inversely proportional to reduced mass, and thus, the smaller the reduced mass of a bond the greater the energy (and frequency) required for vibration. For example, C–H has a smaller reduced mass than C–C and therefore stretching is induced at a higher frequency.

- As the mass of the atom bonded to carbon increases (i.e. C–H to C–C), the reduced mass increases, less energy is absorbed and the wavenumber decreases.

An important take-home message is that the more energy absorbed by the bond the higher the frequency and faster the vibration. As such, IR spectroscopy data (%T vs. wavenumber) gives important information on the presence and characteristics of bonds in a given biomolecule as a result of absorbance.

Basic equipment and components: The basic setup of an IR spectrometer is essentially similar to that of a UV/visible spectrophotometer, with the primary difference lying in the enhanced optical systems employed. An enhanced optical system provides higher resolution necessary for generation of IR spectra in dual-beam instruments. The IR source is an electric heated filament, which emits a continuous range of frequencies (and thus wavelengths and wavenumbers) in the IR region of the electromagnetic spectrum. The filament is commonly a Nernst filament (comprising Zr, Th and Ce oxides) or Globar (silicon carbide), and the emitted beam passes through the sample in a cell (container) of defined path length (typically 10^{-2} cm for solutions), where IR is absorbed by the sample. As with UV/visible spectroscopy, IR measures the intensity reading of a reference (I_0) and a sample (I) to generate %T. As the beam passes through the sample cell, constructed from optically clear sodium chloride (not plastic/glass or quartz), it emerges passing through a series of slits and mirrors (which enhances resolution), ultimately forming a so-called collimated beam. This collimated beam is then passed through a monochromator (prism) that disperses the beam into different wavelengths (light separation) and the detector, which is a thermocouple, converting radiant energy (heat) into an electrical current.

Methodology: The primary measures in IR spectroscopy are %T and wavenumber (that is 1/wavelength). One of the most important methodological aspects of IR spectroscopy is sample preparation. There are a range of approaches for IR spectroscopy, depending on the physical nature of the sample to be analysed (i.e. solid, liquid, gas). A **solid sample** is typically prepared by crushing the sample with a mulling/dispersing agent such as Nujol (mixture of hydrocarbons) and applying a thin layer of the mull to the sample plate/cell. An alternative method is to finely grind dry solid sample with potassium bromide powder (purified salt), which is then crushed in a mechanical die press to generate a translucent potassium bromide disk (the sample plate/cell). A **gas sample** is introduced into a cylindrical sample cell (of inactive IR material, e.g. potassium bromide or sodium chloride) that has a regulated inlet and outlet port to control gas flow into the cell during the experiment. A **liquid sample** is simply introduced between the two plates (of inactive IR material, e.g. potassium bromide) comprising the sample cell, and usually the liquid sample is prepared using an anhydrous solvent such as chloroform as water can destroy the sample plate. Importantly, the chosen method of sample preparation can affect the spectra (position of peak absorptions) and as such, caution is required in data interpretation. In a typical IR spectrometer the sample and reference are run in parallel, being present in the instrument at the same time. The spectrum ($\%T$) is recorded over the range $4000–400\,cm^{-1}$, generating dips as opposed to the absorbance peaks of UV/visible spectra.

Interpretation of IR spectra: While it is generally recognized that there are no hard and fast rules for interpreting an IR spectrum, there are a number of fundamental considerations. Firstly, the intensity peaks of the IR spectrum must be of adequate magnitude, and peaks should be clearly visible and resolved. Other important factors include (i) purity and concentration of sample being analysed, (ii) solvent used for preparation of sample and reference, (iii) properties of sample plate/cell, and (iv) calibration of the instrument. The IR spectra generated can be divided into four principal regions relating to molecular vibrations, distinguished on the basis of wavenumbers (cm^{-1}). These all lie in the IR region of the electromagnetic spectrum and are $4000–2500\,cm^{-1}$, $2500–1900\,cm^{-1}$, $1900–1500\,cm^{-1}$, and $1500\,cm^{-1}$ and below (so-called *fingerprint* region). In some circumstances measures may be limited to any one of these regions (e.g. 4000–2500 is often one of the first measures). **4000–2500 region:** area where bonds to hydrogen usually absorb – that is, low reduced mass of X–H, where X may be C, N, O, S. **2500–1900 region:** area where triple bonds usually absorb – for example C≡X, where X may be C or N. **1900–1500 region:** area where C double bonds usually absorb – for example C=X, where X may be C, N or O (plus N=O). **Fingerprint region:** area where many bonds absorb, particularly single bonds, and complex in interpretation – for example C–O, C–N, C–C, C–Cl.

Due to the complexity in interpretation of data in the fingerprint region it is preferentially used for direct comparison with pure known compounds, to identify those compounds in a sample. However, as illustrated in the table below (Table 5.3), certain compounds will give certain bands in different IR regions, which help with identification. As, for example, compounds such as aldehydes and amides have peaks in the same regions, more in-depth analysis relying on the exact wavenumbers and use of defined tables is required for specific identification.

Table 5.3   Examples of IR peaks for common organic compounds

Compound	Bonds	Peaks identified in IR region
Alcohols	O–H	4000–2500
	C–O	Fingerprint
Aldehydes	C–H	4000–2500
	C=O	1900–1500
Alkanes	C–H	4000–2500
Alkenes	C=C	1900–1500
Alkynes	C–H	4000–2500
	C≡C	2500–1900
Amides	N–H	4000–2500
	C=O	1900–1500
Carboxylic acids	O–H	4000–2500 (stretching)
		Fingerprint (bending)
	C=O	1900–1500
	C–O	Fingerprint
Ketones	C=O	1900–1500
	C–C–C	Fingerprint (stretching)

Another reason for using regions as opposed to exact wavenumbers relates to the fact that readings can be made under different conditions. For example, a compound may give a different reading in one solvent compared with a second solvent, and for convenience there are defined tables which can be used to aid interpretation in each case.

Examples of IR spectra arising from analyses of unknown samples are given in Figure 5.7, including interpretation/identification of particular compounds in the sample.

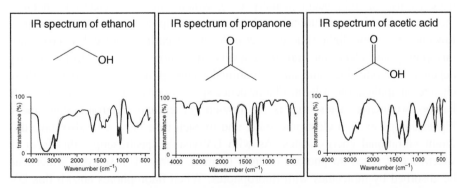

Figure 5.7   Typical IR spectra identifying an alcohol, ketone and carboxylic acid.

Applications: As IR spectroscopy is principally used for structural analyses, this technique has a number of applications in research and industry including quality control. Moreover, modern instruments can relate unknown samples to known spectra stored in a computer database, generating automatic identification of compounds in, often complex, samples. Other specific features of IR spectroscopy have proven useful in specific areas; for example, in dynamic IR spectroscopy where measures are taken over time and many times per second, changes in bonding can monitor chemical reactions (e.g. polymerization). Virtually all modern IR spectrometers are Fourier transform instruments which have a number of advantages over conventional spectrometers. However, Fourier transform infrared (FTIR) spectroscopy is not the same as conventional IR spectroscopy, as it uses an interferometer instead of a monochromator. In FTIR spectroscopy, the measured signal is an interferogram and, after mathematical Fourier transformation, a spectrum is produced which is identical to that generated by conventional IR spectrometers. Interferometers are cheaper to build than monochromators, and measurements from FTIR instruments are quicker, with enhanced sensitivity. Indeed, some would state that FTIR is more versatile than, and superior to, methods such as GC-MS. For this reason FTIR spectroscopy has become popular in forensic toxicological analyses. Other variations of IR spectroscopy include two-dimensional infrared (2DIR); spectroscopy that relies on 2D correlation analysis of IR spectra, which simplifies spectral analysis and improves resolution, finding new avenues for research and other applications.

5.5   Principles and applications of fluorescence spectrofluorimetry

Background and principles: Fluorescence spectrofluorimetry is a type of electromagnetic spectroscopy that measures characteristic fluorescence and generates fluorescent spectra, namely absorbance and emission spectra. This form of spectroscopy uses a light beam (usually UV), which can excite electrons in chromophores (in this case fluorophores) in certain compounds, causing emission of low-energy visible light (fluorescence). As noted earlier, molecules have various energy and electronic transition levels, and fluorescence spectrofluorimetry depends on both electronic and vibrational states of molecules and how these change on application of light. Fluorescence spectroscopy centres on the molecule having two major states; that is the low-energy ground electronic state and a higher-energy excited electronic state, and in each case these will also have different vibrational energy states. As with UV/visible spectroscopy, this form of spectroscopy relies on photons of light (coming from a UV source), which can be absorbed by molecules with fluorophores. When fluorophores absorb this UV light they are excited to a higher-energy electronic state, and on emission (return

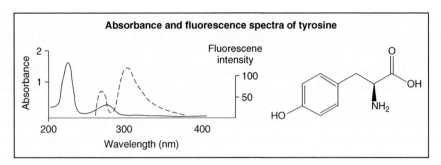

Figure 5.8 **Absorbance and emission spectra of tyrosine.**

to ground electronic state) they cannot simply lose all of their excess energy as heat; rather they emit a portion as fluorescent light in a phenomenon termed *fluorescence*. It is also important to note that when the molecule is excited in this way it reaches one of the vibrational states in the excited electronic state, and some of the light energy initially absorbed is lost in transitions between these vibrational states. As such, the light energy emitted is always of longer wavelength (lower energy) than that absorbed, giving rise to separate *absorbance* and *emission spectra* (see Figure 5.8). Interestingly, by studying the different frequencies of light emitted, the structure of different vibrational levels can be elucidated, giving useful insights into biomolecular structure.

Basic equipment and methodology: Devices that measure fluorescence are called *spectrofluorimeter*s (see Figure 5.9). These instruments comprise a UV source

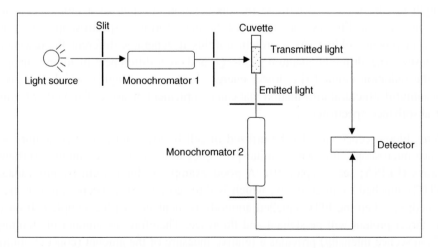

Figure 5.9 **Diagram illustrating setup of a typical spectrofluorimeter.**

that emits light, passing through a monochromator before reaching the cuvette containing the sample and composite fluorophores. While transmitted light passes directly from the sample to the photomultiplier tube (detector), emitted light is detected by a monochromator (which measures different wavelengths). Although emitted fluorescent light passes out of the cuvette in all directions (i.e. cuvette has four clear sides), placing the monochromator at 90° to the incident light (entering the sample cuvette) helps ensure that the signal picked up by this monochromator is emitted light. The detector also picks up the emerging signal from this second monochromator. As such there are two main measures, being intensity of emitted (or transmitted) light as a function of the wavelength of the incident light (excitation spectrum) or emitted light (emission spectrum), or both.

Note: A spectrofluorimeter is quite different from a luminometer (which measures chemiluminescence), and it is important not to confuse the two. Unlike fluorescence, chemiluminescence relies on a chemical reaction between, for example, firefly luciferase and luciferin and molecules in the sample. For measures of chemiluminescence, molecules promoted to an excited state as a result of the reaction return to the ground state with emission of light, which can be measured.

Applications: There are very many applications of spectrofluorimetry in biomedical and chemical research to analyse the presence and features of organic biomolecules. Perhaps the most popular use is in the field of biochemistry in measures of molecule binding, folding and kinetics. For example, if binding of a biomolecule to another protein/receptor causes a change in the 3D structure, the behaviour of the composite fluorophore, reflected in changes in the fluorescence spectrum, can be monitored using spectrofluorimetry. Moreover, providing the fluorophore has a unique location in the biomolecule, changes in fluorescence at a given wavelength can be used to determine the dissociation constant of a biomolecule to protein/receptor. Fluorescence is one of the main forms of spectroscopic analysis used in protein folding. As a protein unfolds it generates different fluorescence intensities (i.e. different quantum yields, Q) and other differences including change in the radiation emitted (i.e. lower energy, longer wavelength), which can give meaningful structural information that can complement measures from other forms of absorbance spectroscopy.

Note: Fluorescence can also be utilized in cell biology and in particular fluorescence microscopy, following similar principles to enzyme-linked immunosorbent assays (ELISA; see Chapter 10). A good example is fluorescein isothiocyanate (FITC) attached to an antibody, which is directed against a specific antigen presented; so when the FITC-labelled antibody is incubated with a sample cell only the target protein will be stained and fluoresce. Therefore the amount of staining (fluorescence intensity) provides a relative measure of the amount (concentration) of the target protein.

Key Points

- At the heart of spectroscopy, spectrometry and spectrophotometry is the interaction of waves of energy (electromagnetic radiation and non-electromagnetic radiation) and matter.

- Spectroscopy relies on the use of electromagnetic radiation to give important structural information regarding biomolecules in a sample.

- Spectrometers allow determination of important parameters including wavelength and intensity, enabling identification, characterization and quantification of given biomolecule(s).

- Dual-beamed spectrophotometers utilize two separate radiation sources, generating two individual beams, which interact with the sample and reference to allow measurements.

- Absorption spectroscopy records electromagnetic radiation (energy) as atoms absorb energy, and conversely emission spectroscopy records energy emitted during atomic transitions.

- UV/visible spectrophotometry utilizes radiation in the UV region of the electromagnetic spectrum to induce electronic transitions.

- A popular use of a UV/vis spectrophotometer is determination of concentration of a biomolecule in a sample using the relationship between concentration and absorbance – the Beer–Lambert law.

- IR spectroscopy uses radiation in the IR region of the electromagnetic spectrum to induce vibrational excitation of bonds in a compound/biomolecule, rather than electronic transitions.

- Measures obtained using an IR spectrometer are $\%T$ and wavenumber, generating an IR spectrum, which gives important structural information about biomolecule(s) in a sample.

- Fluorescence spectrofluorimetry is a type of electromagnetic spectroscopy that generates absorbance and emission fluorescent spectra, giving insights into biomolecular structure.

6 Centrifugation and separation

Bioanalytical chemistry relies on the ability to separate out biomolecules from complex mixtures in biological samples. This chapter considers some of the fundamental methods for separation of molecules and the major components of cells that make up body tissues. The oldest form of bioanalytical separation is centrifugation, which remains a key feature of virtually every laboratory, and this chapter considers the centrifuge and its applications. The first reported use of centrifugation was by Antonin Prandl who used this technique to separate cream from milk in the preparation of butter. Since then, centrifugation has evolved to become one of the most widely used basic separation techniques. The following sections provide a concise and informative outline of the key principles and applications of centrifugation and other key forms of separation technology.

Learning Objectives

- To appreciate the variety of popular methods to separate and isolate biomolecules.

- To understand the core principles underlying centrifugation.

- To describe the features and components of major types of centrifuges.

- To illustrate how major centrifugation methods are utilized for bioanalysis.

- To outline the principles and applications of flow cytometry.

Understanding Bioanalytical Chemistry: Principles and applications   Victor A. Gault and Neville H. McClenaghan
© 2009 John Wiley & Sons, Ltd

6.1   Importance of separation methods to isolate biomolecules

Methods to separate or fractionate biological and biomedical samples lie at the heart of a diverse range of scientific disciplines including biochemistry, cell biology and molecular biology. The choice and mode of separation is important to obtain the best results and avoid artefacts. There are a number of physical and chemical separation technologies routinely used by the bioanalytical chemist that are described in this chapter and various other chapters of this textbook.

Ultrafiltration

The term *ultrafiltration* refers to a type of *membrane filtration*, that is, a method whereby a solution passes through a porous (semi-permeable) membrane under pressure, selectively allowing passage (either diffusion or facilitated diffusion) of certain molecules or ions (see Figure 6.1a). *Diffusion* is a process whereby molecules or ions move from a higher concentration to a lower concentration (concentration gradient), and the rate of diffusion depends on factors such as pressure, concentration and temperature on either side of the membrane. Ultrafiltration can be used for (i) concentrating dissolved biomolecules; (ii) introducing biomolecules from one solution to another; (iii) protein and peptide enrichment and separation (including use of centrifugal concentrator); (iv) sterilizing solutions

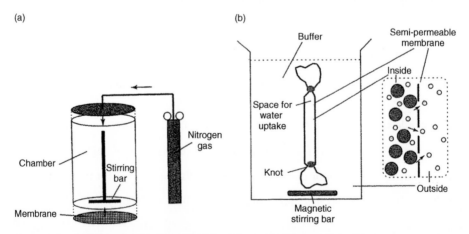

Figure 6.1   Outline of apparatus used in (a) ultrafiltration and (b) dialysis. (Adapted from Sheehan, Physical Biochemistry, 2000.)

(retaining bacteria in the membrane); and (v) removal of salts and impurities from a buffer (using a centrifugal concentrator).

Dialysis

This technique is commonly encountered in laboratory separation chemistry, and relies on the same basic principles as medical dialysis (e.g. kidney dialysis). In essence, this technique involves the use of a semi-permeable (dialysis) tube (or bag) in which a certain solution is placed. This *dialysis tube* is tightly sealed at both ends to retain the solution, before transfer to another solution (e.g. water or buffer) in a container. The solutions are then allowed to *equilibrate*, that is, water will move in or out of the tube and biomolecules will also pass through the semi-permeable tubing by diffusion. Thus, biomolecules at higher concentration in the solution contained in the tubing will move (diffuse) into the solution in the container (see Figure 6.1b). Importantly, larger biomolecules (e.g. proteins, nucleic acids or polysaccharides) typically cannot pass through the membrane as they are larger than the membrane pore size, and thus are retained in the tubing. As such, this technique has commonly been used to remove salts from protein solutions. To enhance diffusion, which is typically slow, it is possible to apply an electrical current to the solution in the container, a technique referred to as *electrodialysis*.

Precipitation

This process describes a chemical reaction in which a solid (insoluble precipitate) is formed in a solution. In its most basic form, *precipitation* may be the generation of a salt by the reaction of an acid plus a base (forming salt plus water). The insoluble *precipitate* will be either more or less dense than the solution in which it forms. If the precipitate is more dense it will typically sink (under gravity) to the bottom of the container, whereas if it is less dense it will typically float in the solution as a suspension. The idea of precipitation is to allow separation, selectively drawing a molecule out of a solution, which may precede centrifugation. When considering biomolecules, there are a number of reagents for precipitation including (i) alcohol; (ii) ammonium sulfate; (iii) acetone; (iv) tricarboxylic acid (TCA); and (v) antibodies (see Chapter 10 for further discussion on immunoglobulin molecules). Precipitation is a fairly crude separation method and may thus precede further more refined separation and subsequent analysis, for example sodium dodecyl sulfate polyacrylamide gel electrophoresis (SDS-PAGE) or chromatography (these techniques are described in more detail in Chapters 7 and 8).

Sedimentation

This process is fundamentally important when considering centrifugation, but does not necessitate the use of a *centrifugal force*. In its most rudimentary form, molecules in solution or particles in suspensions move under gravity to *sediment* (settle) at the bottom of a solution. However, there may be other external forces over that of gravity that can be applied to enhance sedimentation and yield.

6.2   Basic principles underlying centrifugation

Centrifugation is a simple physical method of separation that relies on applying an intense force, called *centrifugal force*, to a sample in order to separate out components from a mixture. During the process of centrifugation, particles are forced through a solution (containing the sample mixture) and are separated out on the basis of their relative resistance to movement through the solution. This technique relies on various properties of the particles and solution. Important physical factors of the particles such as mass, shape/size and density will affect their ability to move under centrifugal force through the solution. The properties of the solution, including temperature, viscosity ('thickness' or resistance to flow), density and composition define the ability of the particles to move through the solution under centrifugal force. These properties and physical factors taken together engender the versatility of this technique in the sedimentation or fractionation of biological samples.

All particles in a solution are subject to the force of gravity, which will eventually (at least theoretically) cause them to sediment. Taken most simply, fast rotation of a solution creates a centrifugal force, which greatly enhances the sedimentation of suspended particles. A useful illustration of the concept of centrifugal force (which comes from the Latin *centrum* 'centre' and *fugere* 'to flee') is the popular event in the Scottish Highland Games called *throwing the hammer*. This is illustrated in Figure 6.2. As shown, the ball of a certain mass (*m*) at the end of the hammer (distance *r* from the shoulder) initially rests on the ground at an angle to the body. Once the competitor starts to spin around on their feet at a given axis, the hammer lifts from the ground and a force (*F*) is generated on the ball at the end of the hammer, and the angle between the ball and the body changes with the velocity (so-called angular velocity or ω).

This basic illustration helps to understand the principles underlying centrifugation and centrifugal force, and this force related to gravity. The movement of any physical matter is governed by Newton's *Laws of Motion*, and when considering centrifugal force it is important to also consider the *centripetal force*. These two form the important action–reaction force pair of circular motion, where the

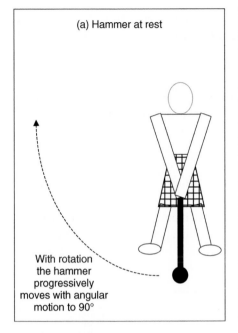

(a) Hammer at rest

With rotation
the hammer
progressively
moves with angular
motion to 90°

(b) Hammer in motion

Figure 6.2 **Diagrammatic illustration of concept of centrifugal force.**

centripetal ('centre seeking') force – the action – is balanced by the centrifugal ('centre fleeing') force – the reaction.

Centrifugal force and relative centrifugal force

If a particle of a given mass is subjected to a centrifugal force it will behave in a certain manner and as detailed earlier it will move through a solution in a defined way, which is given by the following formula:

$$F = m\omega^2 r$$

where F is centrifugal force (on the particle), m is mass of the particle (g), ω is angular velocity (radians per second, rad s^{-1}), r is distance from central axis of rotation (cm).

Given that the particle is usually of fixed mass, the variables angular velocity and radius largely define the centrifugal force on the particle. The angular velocity depends on (and can be altered by) the speed of rotation, and the radius is predefined before centrifugation begins. Both angular velocity and the radius are dependent on the specific rotor used.

As gravity also plays a role in sedimentation, this relates to the centrifugal force in a way defined by the following formula:

$$RCF = 1.118^{-5}(\text{rpm})^2 r$$

where RCF is relative centrifugal force (on the particle), 1.18^{-5} is an empirical constant (related to gravity), rpm is number of revolutions of the particle per minute (revolutions per minute, $r\ \text{min}^{-1}$), r is distance from central axis of rotation (cm).

The relative centrifugal force (RCF) value is thus related to rpm – a popular way of reporting the speed of a given centrifuge with a particular rotor. However, rpm is not a very scientific way of expressing the experimental conditions used, as it relates to the use of a specific rotor. So when a different rotor is used with the same rpm any differences in angular velocity and radius

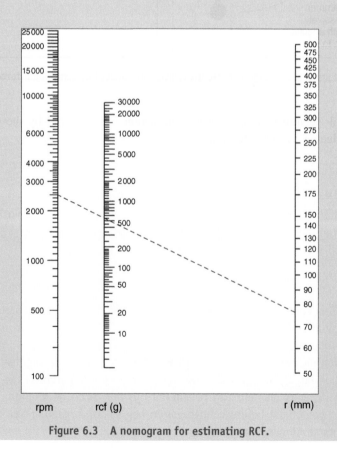

Figure 6.3   A nomogram for estimating RCF.

will alter the centrifugal force. Thus, reporting an rpm value without specific details of the rotor does not allow a person using a different centrifuge to recreate the same experimental conditions. As such, RCF is expressed as a number multiplied by gravity (g), giving a value independent of the rotor used.

The RCF value may also be determined using a nomogram, which is a conversion table allowing the use of rpm and radius values to estimate this force. A typical nomogram is given in Figure 6.3. Using a radius value of 77 mm and knowing that the rpm is 2500, the RCF value is estimated as $550 \times g$ (or $550g$).

The rate of movement of the particle is given by the sedimentation coefficient (s), which depends on the mass and buoyancy of the particle in solution, and the goal of many centrifugation experiments is to measure s (expressed in Svedberg units, S). Different particles, biomolecules and cellular components have defined s values that can relate to molecular weight or size (e.g. for human haemoglobin $s = 4.5$ S, at 20 °C in water). In general terms, the greater the molecular weight/size the higher the s value (see Table 6.1).

Table 6.1 Examples of sedimentation coefficients (s values) for common proteins

Protein	Molecular weight	Sedimentation coeffecient
Cytochrome c (bovine heart)	13 370	1.17
Haemoglobin (human)	64 500	4.5
Catalase (horse liver)	247 500	11.3
Urease (jack bean)	482 700	18.6
Tobacco mosaic virus	40 590 000	198

6.3 Features and components of major types of centrifuge

The principles of centrifugation outlined above are applied in a laboratory context through use of a piece of equipment called a *centrifuge*. A centrifuge acts to sediment (or pellet) particulate matter or cells from a biological solution or suspension by applying a centrifugal force to the sample. The principal components and major types of centrifuge used for a variety of (specialized) purposes are considered below.

Key components and safety aspects

The main components of any centrifuge are an electric motor and drive shaft (together referred to as a *spindle*) and a rotor to hold samples in containers, typically centrifuge tubes. An electric motor drives the centrifuge and applies acceleration to the rotor, attaining an appropriate number of revolutions per minute (rpm) and associated centrifugal (g) force (see Figure 6.4).

To give an idea of the impact of centrifugal force, even the lowest speed centrifuges can reach $6000\,g$; top pilots can handle around $9\,g$, and even brief exposure of humans to $100\,g$ can be fatal. Size and power of a motor depends on purpose and type of centrifuge used (see below). Centrifuges use various types of rotor, including the commonly encountered swinging bucket, fixed angle and vertical rotors illustrated in Figure 6.5.

Modern rotors are often made from carbon fibre, superseding older aluminium or titanium rotors that tended to corrode and fatigue. Carbon fibre rotors are also lighter, thus allowing the samples to reach the optimum acceleration in a shorter period of time. Fixed angle rotors are useful for differential centrifugation (pelleting), whereas swinging bucket or vertical rotors are commonly used for density gradient centrifugation.

When using a centrifuge there are a number of considerations with regard to maintenance and safety. It is important to check the rotor for damage and cleanliness before use, and ensure that the rotor is properly installed in the centrifuge and the equipment is operational. Obvious care should be taken with respect to standard operating procedure, as centrifuge rotors can reach very high velocities and in so doing generate substantial amounts of kinetic energy. If not correctly installed on the spindle or properly balanced, a rotor may leave its circular path

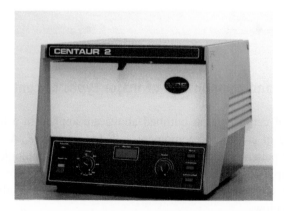

Figure 6.4   Photograph of a low-speed centrifuge.

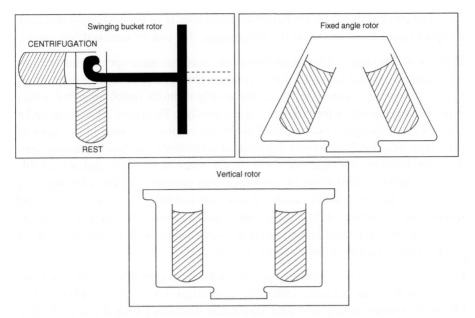

Figure 6.5   **Diagram of swinging bucket, fixed angle and vertical rotors.**

of motion, damaging the centrifuge, and in some cases may even leave the centrifuge, causing considerable damage to persons and property in its path. Care should also be taken to ensure that the correct type of centrifuge tube is used, and it is also good practice not to leave the centrifuge unattended until it has reached its maximum velocity.

Types of centrifuge

There are three major types of centrifuge commonly encountered in laboratory settings; namely low-speed (clinical) centrifuge, high-speed centrifuge and ultracentrifuge. These have different characteristics and applications, which are briefly considered below.

Low-speed (clinical) centrifuge: As the name suggests, these centrifuges require the lowest power motors to achieve desired velocities, and are routinely used to sediment relatively heavy particles or cells. This class of centrifuge encompasses so-called bench (or bench-top) centrifuges, and is found in virtually every bioanalytical lab (see Figure 6.4). Low-speed centrifuges have capacities of 6000 g and typically operate at room temperature, utilizing either swinging bucket or fixed angle rotors. Sample volumes are commonly in the range of 10–50 ml and a range of specialized tubes handle such volumes, which often slot into adaptors in the rotor. During centrifugation a pellet forms at the bottom of the tube, separated from an upper liquid layer known as the supernatant, that can be removed (decanted).

High-speed centrifuge: These centrifuges are used for more specialized applications requiring higher velocities for separation. This type of centrifuge often offers in-built temperature control (refrigeration) to help maintain sample integrity, which is particularly important for biological specimens (see Figure 6.6). The three major types of rotor (swinging bucket, fixed angle and vertical) are used in this class of centrifuge, selected for purpose. High-speed centrifuges can handle various sample sizes, but typically these are small, in the order of $1-10$ ml, and are capable of up to $50\,000g$. This class of centrifuge also encompasses the microfuge (or minifuge), which is a small bench-top unit that handles low volume $0.5-1.5$ ml specialized microfuge (Eppendorf) tubes, and impressively can deliver up to $12\,000g$. Common uses of high-speed centrifuges include sedimentation of cellular components (organelles) and microorganisms.

Ultracentrifuge: As the name implies, this is the most complex type of centrifuge equipment that can generate incredible centrifugal forces (up to $1\,000\,000\,g$). Given the kinetic energy and heat generation associated with the use of ultracentrifuges, refrigeration is absolutely necessary, and high vacuums are needed to reduce friction. Again, proper maintenance and safety measures are imperative

Figure 6.6   **Photograph of a high-speed bench centrifuge.**

when using such a powerful machine. This class of centrifuge can be further sub-divided into *preparative* and *analytical* instruments. Preparative models are used for separation and purification, whereas analytical models offer the ability to perform physicochemical measurements such as sedimentation velocity and equilibrium. Indeed, analytical ultracentrifuges with inbuilt optical detection systems can give information on the gross shape of, and conformational changes in, biomolecules, and size distribution within samples, as well as equilibrium constants.

6.4   Major centrifugation methods for bioanalysis

There are two major forms of centrifugation commonly encountered; namely *differential centrifugation* and *density gradient centrifugation*. Furthermore, as indicated earlier, centrifugation can be used in both preparative and analytical modes, providing powerful tools for bioanalysis.

Differential centrifugation

This technique is sometimes referred to as *differential sedimentation*, and is essentially a process of successive centrifugation (single or repeated steps) with increasing centrifugal force (g). Separation is predominantly dependent on particle mass and size, where heavier particles or cells settle first at lower g values (e.g. intact cells can sediment at around $800\,g$). However, in many cases, differential centrifugation is used to separate out intracellular matter, and thus this method is important for so-called subcellular fractionation (see Figure 6.7).

During subcellular fractionation, various markers can be used as a quality control measure, giving an assessment of the quality of separation of individual fractions; for example, DNA can be used as a marker for the step sedimenting nuclei, while the enzyme succinate dehydrogenase can be used as a marker for the step sedimenting mitochondria. Of course, to obtain a pure organelle fraction from differential centrifugation is virtually impossible.

Density gradient centrifugation

This type of centrifugation can go some way to addressing the main limitation of differential centrifugation, in that it allows preparation of homogeneous organelle fractions and facilitates detailed analysis of cell organelles and function. As the name implies, density gradient centrifugation utilizes a specific medium that gradually increases in density from top to bottom of a centrifuge tube. This

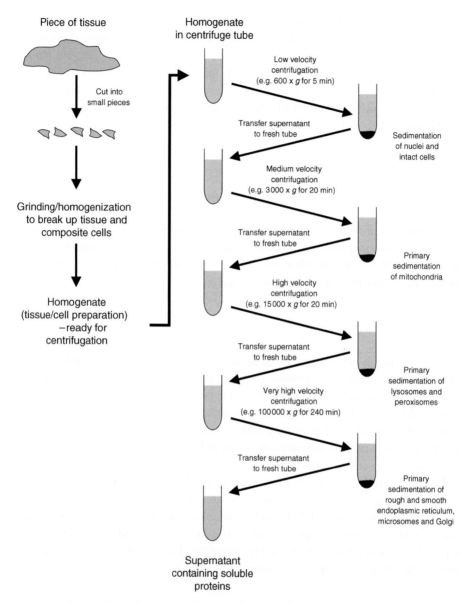

Figure 6.7   **Diagrammatic illustration of differential centrifugation.**

means that under centrifugal force, particles will move through the medium and density gradient and stop (are suspended) at a point in which the density of the particle equals the density of the surrounding medium. The medium used depends on the desired outcome, with four general categories: (i) alkali metal

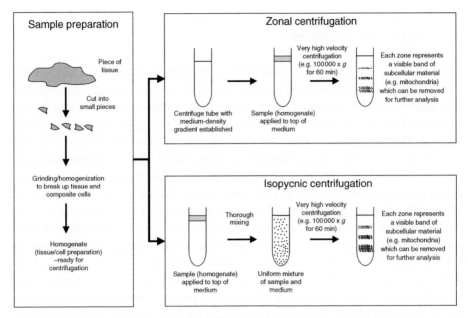

Figure 6.8   **Diagrammatic illustration of zonal and isopycnic centrifugation.**

salts (e.g. caesium chloride); (ii) neutral water-soluble molecules (e.g. sucrose); (iii) hydrophilic macromolecules (e.g. dextran); and (iv) synthetic molecules (e.g. methyl glucamine salt of tri-iodobenzoic acid). The density gradient can be established in the tube either before (zonal) or during (isopycnic) centrifugation. Both *zonal* and *isopycnic* centrifugation are illustrated in Figure 6.8.

For zonal density gradient centrifugation there are a number of media that can be used, however the most common of these is sucrose; whereas the most common medium for isopycnic density gradient centrifugation is caesium chloride ($CsCl_2$).

Preparative versus analytical centrifugation

While preparative centrifugation is relatively limited for direct bioanalysis, there are many virtues of analytical centrifugation, largely based on its capability to use ultracentrifugation to generate actual bioanalytical data. For example, an analytical ultracentrifuge can detect and measure the movement of a particle, yielding data on the sedimentation velocity. Through the use of equations (including the Svedberg equation) incorporating values of sedimentation velocity, a calculation of molecular weight can be made, thus assisting identification and characterization of biomolecules comprising a sample.

6.5   Flow cytometry: principles and applications of this core method of separation

Flow cytometry has emerged as a useful tool for functional analysis at the cellular level, allowing measurements to be made on single cells as they pass through the flow cytometer. This technique builds upon earlier technologies developed by the American Engineer Wallace Coulter, for counting of particulate matter or cells in solution. Coulter reportedly first used this technique to examine plankton particles in seawater, and as they passed through an orifice there were momentary changes (pulses) in impedance (electrical current), which could be counted and expressed as pulses per unit seawater. Later, this technology was adapted to automatically count blood cells (*Coulter counter*) and led to the development of the more advanced method of flow cytometry. In the first instance, the principal application of flow cytometry was in *cell sorting*, but since then it has evolved to encompass a diverse range of measures of physical and biochemical cell parameters. In fact, flow cytometry can, at least theoretically, be used to analyse any cell parameter, providing there is a specific measurable physical property or available biochemical tracer. Examples of targets that can be analysed by flow cytometry include cellular receptors (e.g. membranous G-protein-coupled receptors (GPCRs)), proteins, DNA and RNA.

Principles of flow cytometry

In flow cytometry, measurements are made as cells or particles suspended in a fluid pass through the apparatus in single file. This allows the characterization and measurement (multiparametric) of functional aspects of single cells or particles, and electrical or mechanical sorting of cells/particles into distinct populations. In order to understand the basis of flow cytometry it is important to consider the key components comprising a flow cytometer (see Figure 6.9).

Cells or particles in a sample (in suspension) are passed into a narrow *quartz tube*. This quartz tube is itself housed in, or surrounded by, a sheath, in which circulates a buffer called the *sheath fluid*. Sheath fluid helps concentrate the cells as they pass downwards and are funnelled through the tapered tube into the *injector*. As the internal diameter of the injector narrows, this allows the cells to flow onwards in single file into the flow cell. The flow cell is located in the centre of an optical (and/or electronic) detection system which combines *laser-induced fluorimetry* and *particle light scattering*. That is, when a beam of laser light is directed onto the fluid streaming through the flow cell at the interrogation point, each suspended particle/cell is hit by and scatters light. Moreover, fluorophores in the biomolecules can be excited and emit light (at lower frequency than the

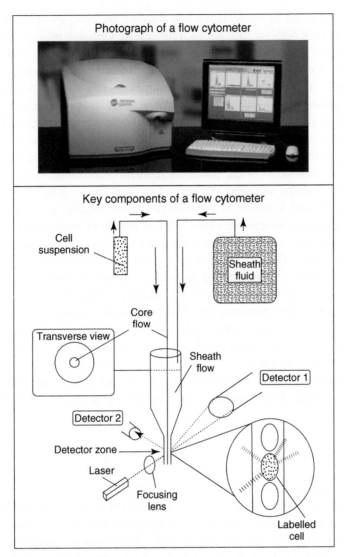

Figure 6.9 Photograph and schematic highlighting major components of a flow cytometer. (Adapted from Sheehan, Physical Biochemistry, 2000.)

source); so, a combination of scattered and fluorescent light is filtered and hits *dichroic* mirrors, which isolate particular wavelength bands. These emitted light signals can be picked up by the *detectors* (photomultiplier tubes) positioned at low angles (e.g. 0.5–10°) to the *incident beam* (so-called forward scatter (FSC)) or higher angles (including 90°; so-called side (or right-angle) scatter (SSC)) and digitized for computer analysis, where information is presented as a histogram (single dimension) or in 2D or even 3D dot-plot formats. By analysis of fluctuations in brightness at each detector (one for each fluorescent emission peak), various types of physical and chemical information about each individual particle/cell can be extrapolated. Common characteristics that can be determined include cell size (and volume), intracellular cytoplasmic complexity, DNA or RNA content, and numerous membrane-bound and intracellular proteins. It is important to note that FSC and SSC give different types of information, so where low-angle FSC may give general measures of particle/cell size, SSC gives an indication of the inner complexity of the particle/cell (e.g. amount/type of intracellular granules).

To give an indication of the power of modern instruments, several thousand particles/cells can be analysed per second in 'real time', also allowing effective separation and isolation of particles/cells with specific properties. Indeed, modern instruments incorporate multiple lasers and fluorescence detectors, allowing use of multiple antibodies, and may even take pictures and allow analysis of individual cells. There has also been a progressive enhancement of available reagents (e.g. fluorescently labelled antibodies), computer analysis software and databases, increasing the versatility of this technique. Specific protocols exist for diagnostic and clinical purposes, and flow cytometers are now emerging as core tools in a number of disciplines, with common applications including determination of: membrane fluidity/permeability; intracellular pH; intracellular ion flux; enzyme and mitochondrial activity (including membrane potential); cell cycle status; protein expression; DNA/RNA expression; and cellular integrity and viability. In particular, flow cytometry has now become a popular tool for clinical haematology.

Major uses of flow cytometry in clinical haematology

Flow cytometry has been exploited for assessment of a number of specimens including blood, bone marrow (where most blood cells are produced), cerebrospinal fluid, urine and solid tissues (e.g. lymph nodes). Combining the use of antibodies, directed against, for example, leukaemia (tumoural) cells, with flow cytometry provides a powerful and rapid means of immunophenotyping, in this case allowing identification of leukaemia cells from various sources (e.g. blood or bone marrow) in a few hours. In addition to this, flow cytometry has found other key applications in blood analyses. Some important illustrations are as follows: blood banking (e.g. measuring erythrocyte surface antigens); genetic disorders (e.g. leukocyte adhesion deficiency, measuring CD11/CD18 complex); haematology (e.g. reticulocyte enumeration, measuring RNA); immunology (e.g. histocompatibility crossmatching, measuring IgG and IgM); and oncology (e.g. S phase of cell cycle, measuring DNA). These are only a few of the very many analyses that can be conducted on erythrocytes (red blood cells), leukocytes (white blood cells), platelets, and other blood soluble biomolecules, using flow cytometry. Indeed, the introduction of smaller and less-expensive instruments means that more routine clinical laboratories can utilize flow cytometry in the diagnosis and management of disease.

Fluorescent-activated cell sorting (FACS)

A major application of flow cytometry is in so-called fluorescent-activated cell sorting (FACS) analysis. When configured for this application, flow cytometers become sorting instruments; but FACS itself does not refer to cell sorting, but rather a trademark method developed by the company *Becton Dickinson*. In essence, FACS is a specialized form of flow cytometry allowing sorting of a heterogeneous (mixed) population of cells in biological samples. In addition to physical separation, this method allows measurement of fluorescent signals from individual cells. The instrument selectively applies a specific charge to a chosen type of cell, achieving physical separation of the subpopulation, and as the cells flow through they are deflected, allowing high-speed (thousands of cells per second) collection in separate tubes.

Other diverse applications of flow cytometry

Given the obvious power of flow cytometry, this technology continues to find new applications in diverse fields of study including: immunology, marine biology, microbiology, medicine and molecular biology, pathology and plant biology. The use of *fluorescence-tagged antibodies* (e.g. green fluorescent protein) has proven to be key to the widening use of this technique in medicine and molecular biology, allowing identification of specific characteristics of cells under analysis, yielding information of diagnostic importance. Other auto-fluorescent properties of cells (e.g. in some photosynthetic marine plankton) have provided new applications of flow cytometry in the characterization of abundance and populations of cells. Flow cytometry can also be used in conjunction with other microbiological methods such as *bacterial display* to indicate cell surface-displayed protein variants in a mixed population of bacterial cells. Indeed, modern flow cytometry enables single or multiple microbe detection (bacteria, fungi, parasites, viruses and yeasts) based on their individual cytometric parameters or fluorophores. Importantly, these approaches also allow assessments of susceptibility of cells to cytotoxic effects of antimicrobial drugs, giving clues as to the responses of heterogeneous populations of microbes to a given antimicrobial treatment regimen.

Key Points

- Bioanalytical chemistry relies on the ability to separate out biomolecules from complex mixtures utilizing physical and chemical separation technologies.

- Commonly encountered separation techniques include ultrafiltration, dialysis, precipitation and sedimentation.

- Centrifugation is a simple physical method of separation that relies on applying a centrifugal force to a sample in order to separate out components from a mixture.

- Important physical factors such as mass, shape/size and density will affect the ability of particles to move under centrifugal force through a solution.

- Relative centrifugal force (RCF) is related to revolutions per minute (rpm) – a means of reporting the speed of a given centrifuge – where RCF is expressed as a number multiplied by gravity (g).

- The main components of any centrifuge are an electric motor and drive shaft (together referred to as a *spindle*) and a rotor to hold samples in containers, typically centrifuge tubes.

- There are three major types of centrifuge commonly encountered in laboratory settings; namely low-speed (clinical) centrifuge, high-speed centrifuge and ultracentrifuge.

- Differential centrifugation is a process of successive sedimentation (single or repeated steps) with increasing centrifugal force (g).

- Density gradient centrifugation utilizes centrifugal force to drive particles through a specific medium where they stop at a point where the density of the particle equals that of the medium.

- Flow cytometry allows characterization and measures of functional aspects of single cells or particles and electrical or mechanical sorting of cells/particles into distinct populations.

7 Chromatography of biomolecules

Earlier chapters have considered several of the major separation techniques used in bioanalysis. This chapter focuses on chromatography, one of the oldest and most fundamental approaches to separating biomolecules from complex mixtures. Chromatographic analysis was first reported in the early 1900s by a Russian botanist, Mikhail Tswett, who employed the technique to separate coloured components of leaves using a simple setup comprising a column containing a few basic chemicals. The term *chromatography* was coined from the Greek words *chroma* meaning 'colour' and *graphia* 'writing', in other words 'colour writing'. The following sections aim to provide a concise and informative outline of the key principles and uses of this important separation technology.

Learning Objectives

- To appreciate the importance of chromatography for separation/identification of biomolecules.

- To describe and explain the principles, types and modes of chromatography.

- To outline key applications of chromatography in life and health sciences.

- To demonstrate knowledge of high-performance chromatography and related technologies.

- To be aware of additional state-of-the art chromatography techniques.

Understanding Bioanalytical Chemistry: Principles and applications   Victor A. Gault and Neville H. McClenaghan
© 2009 John Wiley & Sons, Ltd

7.1   Chromatography: a key method for separation and identification of biomolecules

Chromatography is the most widely used separation technique in bioanalytical chemistry. There are several different modes of chromatography that are based on different physical and chemical attributes of biomolecules, which include separation based on: ionic charge, solute partitioning, molecular size, and adsorption properties. Subsequent sections provide more information on the different modes of chromatography and their application.

The International Union of Pure and Applied Chemistry (IUPAC) define chromatography as a physical method of separation in which the components to be separated are distributed between two distinct phases. The first of these phases is called the *stationary phase* and the other is called the *mobile phase*. Basically, in chromatography, a sample containing a mixture of biomolecules is placed onto a stationary phase (which may be either solid or liquid). A mobile phase (which may be either liquid or gas) is then passed through or over the stationary phase, which causes components in the sample to move along the stationary phase. The diverse physical and chemical properties of individual biomolecules within the sample allow components to separate out because they migrate along the stationary phase at different rates. This migration or separation process is referred to as *elution*, a term you will come across when considering other bioanalytical techniques.

There are two major classifications of chromatography according to whether the separation takes place on a *planar surface* (i.e. flat sheet, e.g. paper or thin-layer

Table 7.1   Basic overview of fundamental chromatography techniques

Chromatography method by classification	Stationary phase	Mobile phase	Separation mechanism
Planar			
Paper	Paper	Liquid	Partition
Thin-layer	Various	Liquid	Adsorption
Column			
Ion-exchange	Resin/matrix	Liquid	Ionic charge
Size-exclusion	Porous resin	Liquid	Size of molecule
High-performance liquid	Various	Liquid	Partition/adsorption
Affinity	Resin/matrix	Liquid	Bioselective adsorption
Gas–liquid	Liquid	Gas	Partition
Gas–solid	Solid	Gas	Adsorption

chromatography) or in a *column* (e.g. ion-exchange chromatography). In addition, these can be further sub-divided into *liquid chromatography* (LC) and *gas chromatography* (see Table 7.1), and the following sections focus on the most popular of the LC techniques.

7.2   Principles, types and modes of chromatography

Separation of biomolecules by chromatography relies on different features of the molecule and chromatography system (i.e. stationary and mobile phases). When considering chromatography, two major principles require attention, namely *retention* and *plate theory*.

Retention and plate theory

Retention: Simply measures the speed at which a biomolecule moves in a given chromatographic system.

Retention is measured as either the *retention time (R_t)* or *retention factor (R_f)*. R_t is used in high-performance liquid chromatography (HPLC) and gas chromatography, whereas R_f is used in paper and thin-layer chromatography. R_f is calculated using the following formula:

$$R_f = \frac{\text{Distance moved by biomolecule along stationary phase}}{\text{Distance moved by mobile phase (eluent or solvent front) along the stationary phase}}$$

Example: Chromatography was used to separate a mixture of biomolecules. The mixture was added to the stationary phase – paper – at a point that was marked as the origin. The mobile phase (eluent or solvent) – ethanol – was then added, and after several hours a number of spots were visible along the paper at distances of (i) 5 cm, (ii) 10 cm and (iii) 15 cm from the origin. During this time the mobile phase moved a total distance of 20 cm (the solvent front). From this, calculate the R_f values of each separated biomolecule:

$$R_f = \frac{\text{Distance moved by biomolecule along stationary phase}}{\text{Distance moved by mobile phase along the stationary phase}}$$

$$R_f \text{ of biomolecule A} = \frac{5}{20} = 0.25$$

$$R_f \text{ of biomolecule B} = \frac{10}{20} = 0.50$$

$$R_f \text{ of biomolecule C} = \frac{15}{20} = 0.75$$

R_f values can be affected by certain factors including temperature, humidity, solvent and type of stationary phase (e.g. paper, alumina, silica). As such, it is important that R_f values are reported along with exact details of solvent used and temperature, to allow reproducibility of the data by other scientists, who may often be in other laboratories.

Typically the solvent used as the mobile phase will be allowed time to move further along the stationary phase than the separated biomolecules. From this, and the formula above, R_f values will normally be calculated as values less than one.

Plate theory: Simply measures the rate of migration of a biomolecule through a stationary phase in a given chromatography system.

This migration is determined by the *distribution ratio* (K_d), otherwise known as *distribution constant* (K_c) that is given by the following formula:

$$K_c = \frac{\text{Concentration of biomolecule in stationary phase}}{\text{Concentration of biomolecule in mobile phase}}$$

Biomolecules with large K_c values will be retained more strongly by the stationary phase than those with smaller K_c values. In other words, as K_c increases it takes longer for solutes to separate.

Various types of chromatography are based on different modes of separation. Figure 7.1 gives a basic overview of the principles and mechanisms underlying the most commonly encountered modes of separation.

Paper chromatography

Background and principles: Paper chromatography is a commonly used method for separation and identification of coloured compounds, including pigments. This technique relies on the separation of biomolecules from a mixture on the basis of *partition*, that is, difference in solubility and hence distribution of a given biomolecule between the stationary and mobile phase.

Methodology: A small droplet of sample containing a mixture of biomolecules is placed (spotted) onto a piece of chromatography paper, and this point is marked

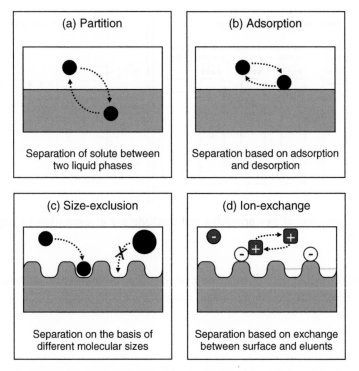

Figure 7.1 **Major modes of chromatographic separation: (a) partition, (b) adsorption, (c) size-exclusion and (d) ion-exchange.**

as the origin (see Figure 7.2). The spotted chromatography paper is allowed to air-dry before transfer into a chromatography tank. The tank, usually made of glass, contains a small volume of solvent, and after the paper is positioned the tank is sealed. Chromatographic separation can be monitored visually by following the movement of the solvent up the paper (stationary phase). The solvent moves by capillary action resulting from the attraction of solvent molecules to the paper and one another, slowly dragging the solvent up the paper. This creates a *solvent front* that can be observed and measured. As the solvent moves up the paper it reaches the origin of the spotted sample(s) and carries along constituent biomolecules, which are subsequently dissolved in the solvent. The constituents of the sample are carried along at different rates, as they are not equally soluble in the solvent, and each has a different degree of attraction to the paper. The chromatography paper is removed from the tank before the solvent front reaches the top of the paper, and allowed to dry. As the spots are usually invisible to the naked eye, the paper (the *chromatogram*) needs to be developed by treatment with chemicals or ultraviolet light, and R_f values calculated. Examples of popular chemical reagents

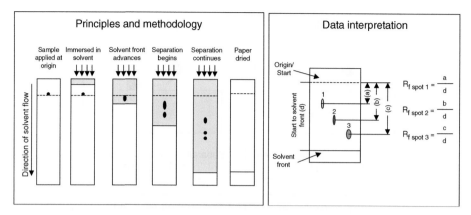

Figure 7.2 Overview of key principles and methodology of separation by paper chromatography.

for visualization and detection are ninhydrin (used for amino acids), rhodamine B (used for lipids) and aniline phthalate (used for carbohydrates).

Thin-layer chromatography

Background and principles: Thin-layer chromatography is the other most commonly used form of planar chromatography and uses a very similar experimental approach to paper chromatography. The principal difference is that this technique relies on the separation of biomolecules from a mixture on the basis of partition and/or adsorption. There is a distinct difference between the process of *ad*sorption and *ab*sorption, and they are not interchangeable terms! Whereas molecules that are *ab*sorbed are 'taken up into', those that are *ad*sorbed 'stick to' a surface. So, in thin-layer chromatography, the mobile phase is adsorbed (sticks to) and subsequently moves along the stationary phase. The stationary phase consists of an adsorbent (sticky) layer on a flat plate or sheet. The most commonly encountered adsorbent layers comprise silica gel, alumina (not aluminium) or cellulose, while popular solvents include hexane, acetone and alcohol.

Methodology: As with paper chromatography, in thin-layer chromatography a small droplet of sample containing a mixture of biomolecules is placed (spotted) onto the adsorbent surface and the origin clearly marked. Separation follows a similar process, however thin-layer chromatography has a number of advantages including: enhanced speed of separation, greater resolution, a wider range of adsorbents, and easier isolation of spots from the chromatogram (Figure 7.3).

Figure 7.3   Photograph illustrating a chromatography tank used in thin-layer chromatography.

As the spots are usually invisible to the naked eye, the paper (the chromatogram) needs to be developed by treatment with chemicals or ultraviolet light.

Size-exclusion chromatography

Background and principles: Size-exclusion chromatography is also known by a number of other names including gel-filtration, gel-permeation or molecular-sieve chromatography. Separation of biomolecules is based on their relative molecular size. In size-exclusion chromatography a column is filled with a substance known as a *gel*, which is the stationary phase. This gel contains a number of *pores* or *holes* of a defined size that is dependent on the nature of the particles making up the gel (see Table 7.2). The gel is often selected on the basis of the molecular weight of the biomolecules to be identified or separated from a sample, and each gel has a characteristic *exclusion limit*. Thus the gel acts as a molecular sieve, retaining molecules up to the exclusion limit and allowing those that exceed this limit (larger molecules) to readily pass through. Therefore, in size-exclusion chromatography, molecules are eluted in order of decreasing size. Importantly, biomolecules are separated based on their size and configuration rather than simply their molecular weight; however, in many cases these are closely correlated. The mobile phase consists of solvents that may be organic or aqueous.

Methodology: Prior to experimentation a gel is selected on the basis of its *fractionation range* and the molecular size of biomolecules to be fractionated (Figure 7.4a). As gels are usually supplied in dehydrated form, it is necessary to add water to allow the gel to rehydrate (swell) before use. This may take hours to days,

Table 7.2   Common types of gels used for size-exclusion chromatography

Gel		Fractionation range (Da)
Dextran (Sephadex)		
Example	G-10	<700
	G-50	1500–30 000
	G-200	5000–600 000
Agarose (Sepharose)		
Example	6B	10 000–4 000 000
	4B	60 000–20 000 000
	2B	70 000–40 000 000
Polyacrylamide (Biogel)		
Example	P-2	100–1800
	P-100	5000–100 000
	P-300	60 000–400 000

depending on the chosen gel. Once the gel has rehydrated to form a 'gel slurry', it must be *defined* – a process where small impurities and dissolved gases are removed to enhance separation. The gel slurry is then poured into a column to tightly *pack* the hollow tube. Solvent is then passed through the column to allow equilibration before addition of solvent containing the known standards. The standard solution containing biomolecule(s) of known molecular weight then passes through the column and biomolecules are eluted based on their molecular weight, where the largest is eluted first. Eluent which has passed through the column is collected at various time intervals, which correspond to elution volume, and each of these aliquots is subjected to further analysis by UV spectrophotometry. From this profile of UV absorbance versus elution volume (which corresponds to time), a graph, which relates molecular weight to elution volume, is prepared representing a standard curve. After this, the column is washed thoroughly with solvent before addition of the sample. As the sample dissolved in the solvent passes through the column, molecule(s) of interest are temporarily retained as they diffuse in and out of the gel, while the larger molecules quickly pass through (are eluted faster). Molecules will elute at certain volumes, which can be observed from UV spectrophotometric analysis, and the molecular mass determined from a standard curve using the elution volume (Figure 7.4b).

Ion-exchange chromatography

Background and principles: Ion-exchange chromatography is a form of adsorption chromatography that relies on the ionic and electrostatic properties of charged

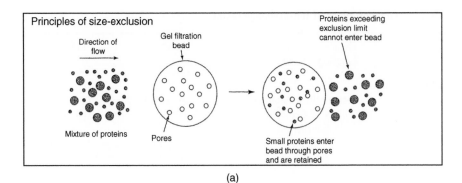

(a)

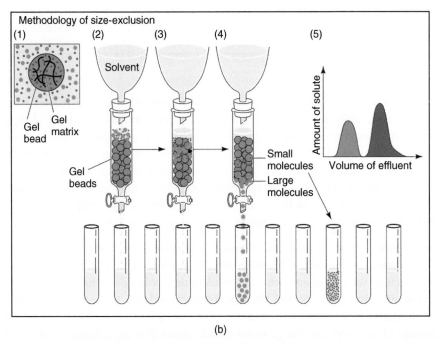

(b)

Figure 7.4 Diagram illustrating the (a) general principles and (b) methodology of size-exclusion chromatography. (Sources: (a) Adapted from Sheehan, *Physical Biochemistry*, 2000; (b) From Voet, Voet & Pratt *Fundamentals of Biochemistry, 2nd edn*, © 2006 Voet, Voet & Pratt; reprinted with permission of John Wiley & Sons, Inc.)

biomolecules. This is perhaps the most conceptually confusing form of chromatography. The separation of biomolecules by this method relies on the exchange of ions between a charged stationary phase and mobile phase of the opposite charge. In ion-exchange chromatography a column is filled with a substance known as a *resin*, which acts as the stationary phase. This resin incorporates chemically bonded functional charged groups with either a negative or positive charge. An

Table 7.3   Common types of ion-exchange resins

Type	Chemically bonded functional group	Resin/matrix	Trade name
Anion exchanger			
Strong base	Tetramethylammonium	Polystyrene	Dowex AG 1
	Diethyl-(2-hydroxyl-propyl)-aminoethyl	Dextran	QAE-Sephadex
Weak base	Tertiary amine	Polystyrene	Dowex AG 3
	Diethylaminoethyl	Dextran	DEAE-Sephadex
Cation exchanger			
Strong acid	Sulfonic acid	Polystyrene	Dowex AG 50
	Sulfopropyl	Dextran	SP-Sephadex
Weak acid	Carboxylic acid	Acrylic	BioRex 70
	Carboxymethyl	Sephacel	CM-Sephacel

anion exchanger comprises a resin containing positively charged (basic) functional groups, while the resin in a *cation exchanger* contains negatively charged (acidic) functional groups. These groups reversibly interact with the mobile phase, attracting biomolecules with the opposite charge. So an anion exchanger attracts negatively charged biomolecules, while a cation exchanger attracts those with a positive charge. Exchangers can be defined as either *strong* or *weak*, depending on the ionizing strength of the chemically bonded functional group. Some of the most important types of ion-exchange resins are given in Table 7.3.

Methodology: Prior to experimentation a resin/matrix is selected on the basis of the nature of the solute biomolecules to be separated. For relatively small biomolecules such as amino acids and nucleotides, polystyrene-based resins have a relatively high capacity and thus are often considered to be the most effective. Capacity is a quantitative measure of the ability of an exchanger to adsorb biomolecules of opposite charge in the mobile phase. For larger biomolecules such as peptides, proteins and nucleic acids, often cellulose-, dextran- or acrylic-based exchangers are selected. Resins are prepared according to the manufacturers instructions and they are usually supplied in slurry (re-hydrated) form ready to be poured into a column. The following steps refer to the preparation and use of a cation exchanger (see Figure 7.5). Once the cation exchange resin (containing negatively charged functional groups) is in the column, it is further prepared for separation. The sample prepared in the loading buffer is then passed through the column. The biomolecules of interest are attracted and *reversibly* bound to the functional groups of the resin while other neutral or negatively charged material dissolved in the buffer passes straight through (is eluted). So the separation has effectively taken place at this point, and all that is now required is to remove

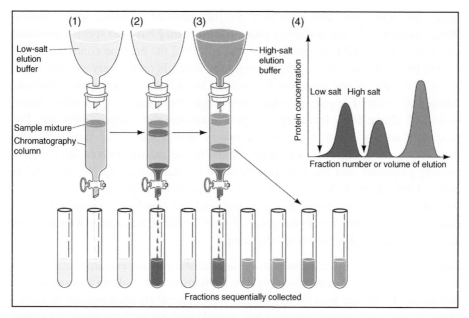

Figure 7.5   Diagrammatic overview of ion-exchange chromatography. (From Voet, Voet & Pratt *Fundamentals of Biochemistry, 2nd edn,* © 2006 Voet, Voet & Pratt; reprinted with permission of John Wiley & Sons, Inc.)

the biomolecules of interest from the resin using an elution buffer with either increased ionic strength or pH. In this case, the elution buffer is a strong base (i.e. with high pH) which displaces bound biomolecules from the resin as they have a weaker charge, and these pass out of the column in the elution buffer for collection and further analysis. Put simply, this is ion bound to resin, ion released from resin.

Affinity chromatography

Background and principles: Affinity chromatography relies on bioselective adsorption, where biomolecules are selected from a mixture on the basis of their unique biological specificity. Separation depends on biomolecules binding to specific sites on the stationary phase. The stationary phase is a resin/matrix to which a fixed functional group is attached. This fixed functional group is able to *irreversibly* bind a specific macromolecule such as an enzyme. This construct thus has three layers: (i) matrix, (ii) fixed functional group, and (iii) specific macromolecule. Sometimes a spacer is placed between the functional group and specific macromolecule to enhance the binding ability of the biomolecule to be separated. Where the stationary phase construct has an enzyme attached to its fixed functional group, this will select and bind to a specific substrate molecule in the mobile phase. Alternatively, the stationary phase construct may have the specific substrate molecule as the fixed functional group that will bind the enzyme contained in the mobile phase. So this type of chromatography depends on the very specific interaction or *affinity* of one molecule for another. In addition to the enzyme-substrate interaction outlined above, affinity chromatography capitalizes on other major biological interactions, including receptor–hormone and antigen–antibody. For successful affinity chromatography the matrix should not adsorb molecules, the fixed functional group must be tightly bound to the matrix without affecting its binding properties and the construct must be resistant to change by the elution buffer. Preparation of the construct is a two-step chemical process: (i) the *activation* of the functional group on the matrix, and (ii) *coupling* (or joining) of the macromolecule to this functional group. Figure 7.6 illustrates a typical method for activation and coupling of an affinity matrix.

Methodology: Prior to experimentation the stationary phase construct is prepared and poured into the column. Notably, the columns used for affinity chromatography tend to be shorter than those used for other column chromatography methods. Once the stationary phase construct has settled (packed) in the column, a loading buffer is passed through to equilibrate the system before addition of loading buffer

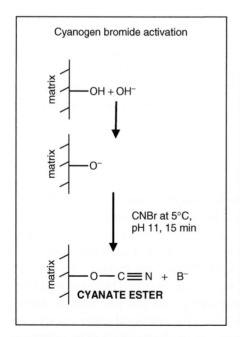
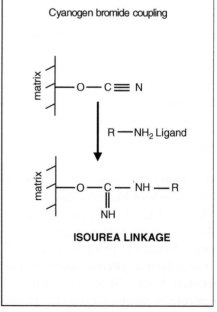

Figure 7.6   Cyanogen bromide activation and coupling of a matrix for affinity chromatography.

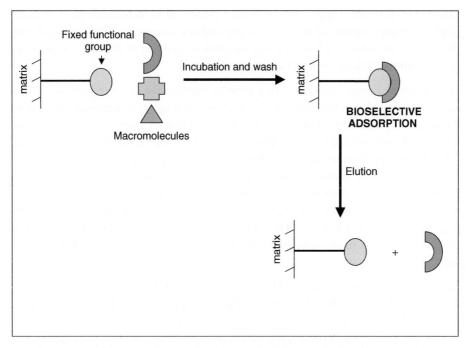

Figure 7.7   **Diagrammatic overview of the principles of affinity chromatography.**

containing the sample of biomolecules. During this process the biomolecules of interest specifically bind to the macromolecule in the construct and thus the separation has taken place and the construct is then washed with loading buffer to remove any impurities. The next step is to remove the biomolecules of interest from the construct using an elution buffer. There are three main types of elution buffer and these are selected for purpose. The first relies on detaching of biomolecules from the construct by weakening binding through decreasing pH or increasing ionic strength. The second, called affinity elution contains a component that displaces the biomolecule from the construct. The final elution buffer contains a chaotropic (denaturing) agent that changes the properties of the construct to weaken the binding to the biomolecule, which is stripped off (Figure 7.7).

7.3   Applications of chromatography in life and health sciences

Paper chromatography: Has largely been superseded by more sophisticated chromatographic methods in bioanalytical laboratories; however this method still remains an important teaching tool to illustrate basic chromatography principles and technique. It still retains value in sample identification using appropriate standards (controls), for example in amino acid analysis.

Thin-layer chromatography: Is commonly employed to determine pigmentation of plants, using extracts of leaves. This method is also used to detect pesticides or insecticides in foods and has been instrumental in analysis of the dye composition of fibres from clothing in forensic investigations.

Size-exclusion chromatography: Is a popular method for the purification of bio-molecules, complementing other forms of chromatography. It is commonly used in the process of *desalting*, that is, removing inorganic salts from a liquid containing the biomolecules of interest. Desalting also refers to the removal of other organic solvents and small molecules, including low molecular weight sugars (e.g. lactose from whey), small organic compounds (e.g. phenol during nucleic acid purifi-cation), fluorescent or radioactive compounds during protein labelling (e.g. fluo-rescein, rhodamine and iodine-125), and detergents used in protein solubilization (e.g. sodium dodecyl sulfate; SDS). The other main application of size-exclusion chromatography is in the determination of molecular weight, because it is simple, inexpensive and fast.

Ion-exchange chromatography: Has a wide range of uses in the food industry, medicine and life and health sciences. This technique has been extensively used in the food industry as a quality control measure and to detect contamination with metals and organic acids. Another important use of this method is in the purifi-cation of blood products such as human albumin, growth factors and enzymes. Perhaps the most common application of this technique is the deionization of water and water softening/purification.

Affinity chromatography: Has a wide number of uses and can be applied to the iso-lation and purification of virtually all biomolecules. Specific applications include nucleic acid purification, protein purification from cell and tissues extracts, and antibody purification from blood serum. There are a number of matrices used for the construct, and some examples of these and their uses are as follows: heparin columns to separate cholesterol lipoproteins, lectin columns to separate carbohy-drate groups, and phenyl boronate columns to separate glycated haemoglobins.

7.4 High-performance liquid chromatography and advanced separation technologies

High-performance liquid chromatography (HPLC) has emerged as one of the most useful tools for the bioanalytical laboratory. This is a powerful stand-alone method that can be used for the purification, separation and quantification of biomolecules. Notably HPLC analysis can be combined with other high-tech methods such as mass spectrometry (MS) or nuclear magnetic resonance (NMR) spectroscopy to enable the identification of biomolecules.

Background and principles: High-performance liquid chromatography is usually abbreviated to HPLC. Historically this method has been referred to as *high-pressure* liquid chromatography, which perhaps gives a better indication of the key principle underlying this technique. In essence, HPLC is used to separate biomolecules from a mixture that is pumped under high pressure in the mobile phase through the stationary phase column. Separation depends on the chemical interactions between the biomolecules in the mobile phase and the column. The first HPLC machine was built by Horvath and Lipsky in Yale University in the late 1960s, and the technique has become increasingly popular through the last few decades due largely to its offering a number of advantages over other classical forms of chromatography.

Advantages of HPLC over classical chromatography methods

- Enhanced resolution and speed of analysis.
- Greatly improved reproducibility.
- Continuing advances in column technology.
- Automated instrumentation.
- Can be scaled up for separation from large sample volumes.
- Columns (stationary phase) can be reused.

HPLC is a very versatile method encompassing all major forms of liquid chromatography. In essence, HPLC is an advanced automated from of partition, adsorption, ion-exchange, size-exclusion, and more recently affinity chromatography.

Basic equipment and components: An HPLC system comprises four key components, namely: (1) solvent pump(s) (delivery of mobile phase); (2) sample injector; (3) HPLC column (stationary phase); and (4) detector linked to recording device. The basic components comprising an HPLC system are illustrated in Figure 7.8.

(1) *Solvent pump(s):* The primary purpose of a solvent pump is to draw mobile phase (solvent) from a reservoir, through a filter, and sufficiently pressurize the solvent to overcome resistance and allow passage through the column into the sample injector. Prior to use, solvents must have all dissolved gases removed (i.e. *degassed* and *purged*) after which they may be drawn into the

Photograph of an HPLC system

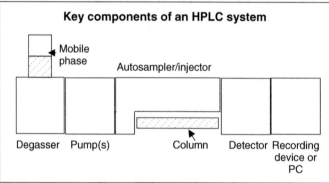

Key components of an HPLC system

Figure 7.8   **Photograph and sketch outlining the principal components of an HPLC system.**

pump by suction. Operation of the pump relies primarily on a piston/piston seal which helps deliver a constant, continuous fixed volume of solvent (*flow rate*) onto the column. An HPLC system may have one or more pumps, and in the case of multiple pumps a *mixer* and *pump flow controller* are required for efficient solvent delivery.

(2) *Sample injector:* The sample injector is located between the solvent pump and the HPLC column, and serves to deliver or inject the sample dissolved in solvent onto the column. The most common form of sample injector is the so-called loop-and-valve injector. Sample is injected from a specialized syringe (manually) or delivered by means of an autosampler (automatically) into a small diameter loop. The loop-and-valve injector allows switching by means of a valve between initial delivery of solvent alone to the column (so-called column equilibration), to subsequent delivery of the sample onto the column.

(3) *HPLC column:* The HPLC column is available commercially as a stainless steel tube pre-filled with fine-diameter packing material. The dimensions of the tube and its packing material define the separation it can achieve. A typical analytical scale HPLC column has an internal diameter of 2–5 mm and is 5–30 cm long. However, it is important to note that column diameters can be <0.5 mm (capillary columns) or over 10 mm, and are selected for purpose. The HPLC column packing material contains particles with pores of various defined sizes and surface areas, characterizing and defining use of the column and the separation it can achieve. One important packing material used in the preparation of HPLC columns is octadecylsilyl-bonded silica (see Figure 7.9). There are two principal modes of HPLC: (i) *normal phase*, and (ii) *reversed phase*. Normal phase HPLC uses non-polar mobile phase (solvent) and a polar stationary phase (column packing material) and using this system, non-polar biomolecules usually elute first while polar biomolecules are retained for longer periods on the column. Conversely, reversed-phase high-performance liquid chromatography (or RP-HPLC) uses a polar mobile phase and non-polar stationary phase, where polar biomolecules elute before non-polar biomolecules.

(4) *Detector linked to recording device:* As the sample passes through the column, biomolecules are separated by their retention in the column and are thus eluted at different time points (retention time). Solvent with dissolved biomolecules passes out of the column into a detector. There are various

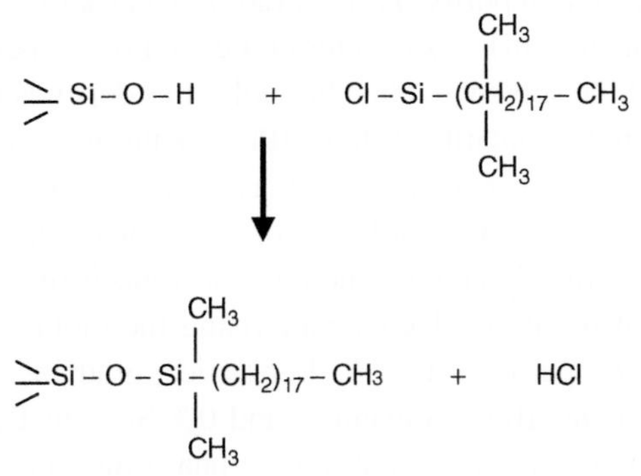

ENDCAPPING: Remaining –OH groups on the silica are blocked with **Trimethylchlorosilane** to prevent these polar groups from interacting with solute, giving poor chromatography

Figure 7.9 **Reaction scheme illustrating the preparation of octadecylsilyl-bonded silica.**

types of detectors including refractive index detector, conductivity detector, fluorimeter, or UV detector. Of these, the UV detector is the most popular due to its versatility. Three commonly used UV wavelengths for detection are (i) 280 nm – specific for aromatic amino acids; (ii) 220 nm – specific for peptide bonds; (iii) 260 nm – for nucleic acids. Data is acquired from the detector by one of a number of alternative recording devices. Traditionally, manual systems employ a strip chart recorder giving a trace, which may either be interpreted directly by the user or facilitated by an integrator that quantifies the data and peak areas. Modern HPLC systems utilize a computer-based system, which in addition to driving the instrumentation can acquire and process data.

Methodology: For HPLC analysis the mobile phase (solvent) is pumped from a reservoir through the sample injector onto the column (stationary phase) and out to the detector linked to a recording device. Prior to experimentation the HPLC column is washed thoroughly by pumping solvent (commonly acetonitrile in water) through to remove any impurities and equilibrate the column for use. As mobile phase passes through the system under high pressure, the sample in solvent is injected into the system and from there quickly passes onto the column. Once the sample is injected into the system this begins the experimental separation and is denoted as time zero ($t = 0$). The principles underlying the separation depend on the type of column used, and the eluted mobile phase passes through the detector where it is collected in sampling tubes.

HPLC mobile phase: The mobile phase is usually an organic solvent and is selected for use on the basis of (i) high purity (HPLC grade), (ii) lack of unwanted reactions with sample or column, and (iii) lack of interference or noise when passing through detector. The ability of the solvent to elute biomolecules depends mainly on solvent strength, which is related to polarity. While HPLC using one solvent that does not change throughout the separation is referred to as *isocratic elution*, for most HPLC separations a blend of two (or more) solvents of different strength are used. HPLC separation depending on this dynamic blending of solvents is referred to as *gradient elution*. The weaker of the two solvents comprising the mobile phase is denoted *Solvent A*, while the stronger is *Solvent B*. In practice, gradient elution starts with a mobile phase comprising 100% Solvent A and 0% Solvent B. Over the course of the desired separation time period the mobile phase introduced into the HPLC system changes with a decreasing percentage composition of Solvent A paralleled with increasing percentage composition of Solvent B. So, at the start of gradient elution the mobile phase entering the HPLC system is 100% Solvent A and 0% Solvent B, and at the end it is 0% Solvent A and 100% Solvent B. The solvent pumps, which are pre-programmed to mix Solvents A and B, control this blending in the desired ratio, over the pre-defined time period of separation.

Interpretation of HPLC data: Data acquired by the detector and processed by a recording device generate a typical HPLC trace of the type shown in Figure 7.10. The HPLC trace diagrammatically presents a number of peaks, each of which correspond to biomolecules (not necessarily individual biomolecules) eluted from the sample. These peaks are interpreted in various ways, particularly in terms of the time at which they emerge (retention time), solvent composition (mixture of Solvents A and B), and the magnitude or size of the peak (*area-under-the-curve*), giving a profile of biomolecules found in a given sample. This profile will also determine the relative purity of a sample and the amount of each component, if required. However, it is important to note that HPLC traces cannot in themselves be used to identify biomolecules, and thus additional measures have to be taken such as the use of a standard curve or coupling of HPLC with complementary bioanalytical instruments such as a mass spectrometer.

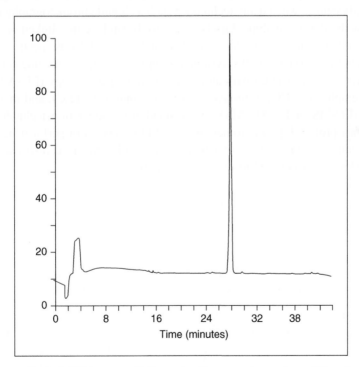

Figure 7.10 **Typical HPLC trace of the naturally occurring gut peptide, glucagon-like peptide-1 (GLP-1).**

Applications of HPLC: Of the bioanalytical separation technologies described in this book, arguably HPLC has the widest range of applications, being adopted for the purpose of clinical, environmental, forensic, industrial, pharmaceutical and research analyses. While there are literally thousands of different applications, a few indicators of how HPLC has been used are as follows: (i) Clinical: quantification of drugs in body fluids; (ii) Environmental: identification of chemicals in drinking water; (iii) Forensic: analysis of textile dyes; (iv) Industrial: stability of compounds in food products; (v) Pharmaceutical: quality control and shelf-life of a synthetic drug product; (vi) Research: separation and isolation of components from natural samples from animals and plants.

7.5   Additional state-of-the-art chromatography techniques

This chapter has considered a range of chromatographic separation techniques. While these techniques have clear stand-alone applications, there are significant inherent advantages when using techniques such as liquid chromatography in combination with other bioanalytical tools. These tools can be either linked or coupled instrumentally to one another. An example of linking is a UV/vis spectrophotometer with a liquid chromatography system. Examples of coupling include combining liquid chromatography (LC) with atomic absorption spectroscopy (LC-AAS), gas chromatography (LC-GC), thin-layer chromatography (LC-TLC), and mass spectrometry (LC-MS or LC-MS/MS). Other stand-alone separation technologies of note are fast protein liquid chromatography (FPLC) and ultra-performance liquid chromatography (UPLC), both of which offer greatly enhanced resolving power when used alone or coupled to other instruments.

Key Points

- Chromatography separates biomolecules on the basis of ionic charge, solute partitioning, molecular size and/or adsorption properties.

- In chromatography there is a physical separation where components are distributed between two distinct phases—the 'stationary phase' and the 'mobile phase'.

- Retention measures the speed at which a biomolecule moves in a given chromatographic system, while plate theory measures the rate of migration of a biomolecule through a stationary phase.

- Paper chromatography relies on the separation of biomolecules in a mixture on the basis of partition, while thin-layer chromatography relies on both partition and/or adsorption.

- Size-exclusion chromatography separates biomolecules on the basis of their relative molecular size, where molecules are eluted in order of decreasing size.

- Ion-exchange chromatography is a form of adsorption chromatography where separation relies on the exchange of ions between a charged stationary phase and mobile phase of the opposite charge.

- Affinity chromatography relies on bioselective adsorption where biomolecules are selected from a mixture on the basis of their unique biological specificity.

- HPLC separates biomolecules from a mixture under high pressure depending on chemical interactions between the biomolecules in the mobile phase and the stationary phase.

- An HPLC system comprises four key components: solvent pump(s) (delivers mobile phase), sample injector, HPLC column (stationary phase), and detector linked to recording device.

- In general, chromatography techniques are both stand-alone or can be coupled with other important bioanalytical tools such as mass spectrometry.

8 Principles and applications of electrophoresis

The principles of electrophoresis and electrophoretic separation are fundamental to many of the most versatile methods of analytical separation. The basis of electrophoresis is the ability to separate charged molecules by means of applying an electrical field. Historically, the development of electrophoresis began with the pioneering work of the Swedish biochemist, Arne Tiselius, who whilst working in The Svedberg's Laboratory at the internationally renowned University of Uppsala, reported the moving boundary method (MBE) of electrophoresis of proteins. Tiselius' work during the 1920s to 1930s resulted in landmark papers paving the way for much-improved methods of electrophoretic analysis. The term *electrophoresis* was coined from the Greek word *phoresis*, which means 'being carried', in other words being carried by an electrical field. The following sections provide a concise and informative outline of the principles and key applications of this important separation technology in life and health sciences.

Learning Objectives

- To outline the principles and theory of electrophoretic separation.
- To describe the major types of electrophoresis.
- To demonstrate knowledge of electrophoresis in practice.

Understanding Bioanalytical Chemistry: Principles and applications   Victor A. Gault and Neville H. McClenaghan
© 2009 John Wiley & Sons, Ltd

- To illustrate key applications of electrophoresis in life and health sciences.
- To give specific examples of advanced electrophoretic separation technologies.

8.1   Principles and theory of electrophoretic separation

Electrophoresis relies on the ability to separate out molecules on the basis of them bearing an electrical charge. This is achieved by applying an electrical field induced between positive (anode) and negative (cathode) electrodes. The movement of ions within electrophoretic apparatus results in two currents, where negative ions (anions) flow from the cathode to anode, and positive ions (cations) move from anode to cathode (see Figure 8.1). This results in an overall net movement of electrons from the cathode to anode, and drives the molecules carrying a particular charge in a given direction through a supporting matrix (e.g. polyacrylamide gels, detailed below). When a mixture of charged biomolecules is placed in an electrical

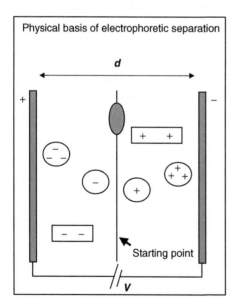
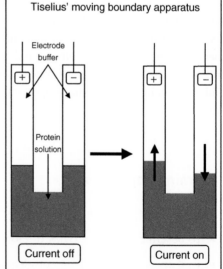

Figure 8.1   Diagrammatic illustration outlining the physical basis of electrophoretic separation and the moving boundary apparatus employed by Tiselius. (Adapted from Sheehan, *Physical Biochemistry*, 2000.)

field with defined *field strength* (E) they are carried towards the electrode of opposite charge. The secret underlying electrophoretic separation in a given experimental system is the fact that different biomolecules have varying physical characteristics (principally net charge, mass and shape) and a driving force called the *Lorentz force* causes them to move at different rates through a given matrix.

A given biomolecule in a mixture will have a net charge (q) that facilitates its movement in an electrical field (E). The velocity (v) at which a biomolecule will move in an electrical field increases with its net charge and the strength of the electrical field. However, the freedom of movement of biomolecules is restricted due to forces of friction (*frictional coefficient, f*) thus decreasing their velocity through the electrical field. The magnitude of the frictional coefficient depends on factors such as the mass and shape of the biomolecule and physical characteristics of the supporting matrix including viscosity and porosity.

In a given electrophoretic system an electrical field is established by applying a voltage (V) to a pair of electrodes (anode and cathode) separated by a particular distance (d). The electrical field strength (E) determines the ability and efficiency of a particular experimental system to separate charged biomolecules. *Electrophoretic mobility* (μ) of a sample depends on the velocity (v) and electrical field strength (E), and biomolecules will migrate based on the ratio of *net charge* (q) to frictional coefficient (f).

In summary, the separation of biomolecules by electrophoresis depends on the experimental system in which they are being separated and the individual properties of biomolecules, namely *net charge, mass* and *shape*. Each of the principles outlined above are summarized by the following important equations:

$$v = Eq/f \quad \text{and} \quad E = vf/q$$
$$E = V/d$$
$$\mu = v/E \quad \text{and} \quad \mu = Eq/Ef = q/f$$

8.2   Major types of electrophoresis

Electrophoresis is a term that describes both a concept and technique or experimental system. There are very many applications of electrophoresis and this technique is still finding new applications in a wide range of scientific disciplines. This section outlines the two major types of electrophoretic experimental systems that are commonly encountered by the bioanalytical chemist.

History of electrophoresis

Tiselius' landmark *moving boundary electrophoresis* (MBE) provided an effective method for separation of molecules in free solution using a U-shaped tank with an electrode at either end. This basic form of electrophoresis was superseded in the 1950s by *zone electrophoresis* (ZE), which relied on the separation of molecules in a solid support (e.g. paper, cellulose). *Paper electrophoresis* was a landmark in separation science in that it paved the way to the now popular two-dimensional (2D) separations used in modern genetic analysis. The most commonly encountered method–*gel electrophoresis* (GE)–was also introduced in the 1950s by Oliver Smithies. Originally, GE was principally performed for bioanalytical analyses, but has since evolved as a major preparative technique to partially purify biomolecules before further characterisation by other advanced technologies including: immunoblotting, mass spectrometry, and molecular techniques such as polymerase chain reaction (PCR) and DNA sequencing. While *starch-block electrophoresis* introduced the concept of *sieving* and the first *discontinuous* buffers, other important developments in the 1960s included Svensson's *isoelectric focusing* (IEF) and Shapiro and colleagues' *sodium dodecyl sulfate-polyacrylamide gel electrophoresis* (SDS-PAGE). Where IEF allowed separation based on surface charge, the now-popular SDS-PAGE permitted separation of biomolecules on the basis of molecular mass. The 1960s also saw the development of *capillary electrophoresis* (CE) by Hjerten, which continues to gain popularity, particularly in the field of clinical chemistry. During the 1970s, *isotachophoresis* represented the first fully instrumental approach to electrophoresis in both analytical (Tachophor) and preparative (Tachofrac) versions, and the introduction of both O'Farrell's *2D mapping* (combining IEF and SDS-PAGE) and *silver staining* techniques greatly increased the sensitivity of electrophoresis. Since the 1980s many of these techniques have evolved to become powerful bioanalytical tools, and novel approaches such as direct interfacing of electrophoretic separation methods with mass spectrometry open new opportunities for researchers to understand fundamental aspects of genomics, proteomics and disease.

A large number of variants of gel electrophoresis are used in bioanalytical analysis to allow separation and characterization of biomolecules, in particular nucleic acids (DNA, RNA) and proteins. The term *gel* refers to the matrix used to separate biomolecules, and in most cases is a cross-linked polymer.

The type of gel is selected on the basis of its composition and porosity, which must be fit for purpose, that is, able to efficiently separate the biomolecules in question. Where the standard methods used to characterize DNA and RNA (200–50 000 bp) are based on agarose gels that have larger but less homogeneous pores, polyacrylamide gels are the best choice when considering separation of small fragments of nucleic acids (up to 500 bp) or proteins (up to 350 000 Da).

Both agarose and polyacrylamide gels are solid but porous matrices, which look and feel like clear jelly (Jell-O). These gels are prepared in different ways but both take the form of a slab (or alternatively column), cast like a jelly in a mould. While agarose gels are relatively easy to prepare, polyacrylamide gels are formed by complex polymerization and chemical cross-linking, and as such are usually purchased pre-cast.

Agarose gel electrophoresis

The study of molecular biology depends on the ability to separate soluble nucleic acid molecules according to their size. As such, agarose GE facilitates important DNA analyses and technologies including PCR, sequence analysis and cloning, and quantification of individual DNA species in a mixture.

Agarose is a component of agar. Agar is an unbranched polysaccharide extracted from the cell walls of some species of seaweed. The term *agar* comes from the Malay word *agar agar*, which means jelly. Agar is a polymer comprising subunits of the sugar galactose, and when dissolved in hot water and cooled agar becomes gelatinous. Agarose is composed of alternating units of galactose and 3,6-anhydrogalactose, and its neutral charge, low chemical complexity and relatively large pore size make it useful for size-separation of large molecules such as DNA (see Figure 8.2). While agarose is used in electrophoretic separation, it is also a common matrix for chromatographic separation (see Chapter 7).

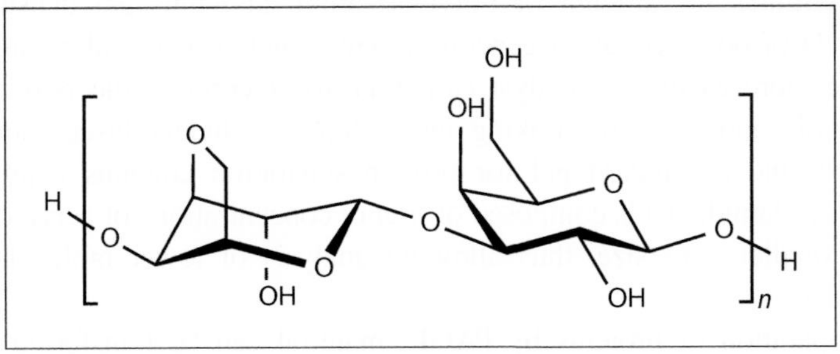

Figure 8.2   **Structure of agarose.**

Agarose gels are prepared by dissolving powdered agarose in hot electrophoresis buffer, and after cooling to ~50 °C the solution is poured into a 'mould', that is, between two glass plates clamped together with spacers. A plastic 'comb' is used to generate 'wells' (indentations) in the gel as it sets. These wells are the points in which the samples are loaded for separation, and represent the point of origin. The mobility of nucleic acids through the agarose during electrophoresis is dependent upon their molecular size and conformation, and the concentration of agarose comprising the gel (typically 0.3–2.0%). Bromophenol blue is a popular tracking dye, that is, a dye that moves through the gel in the same direction, but slightly ahead of the nucleic acids. This allows the operator to monitor movement of the dye and nucleic acids through the gel, thus following the progress of separation. After separation there are various modes of visualization of the material in the gel, which will be discussed later in this chapter.

The molecular biology method *Southern blotting*, named after the British biologist Edwin Southern, is an enhanced form of agarose GE marking specific DNA sequences. Blotting involves the transfer of material from the gel onto a membrane (e.g. nitrocellulose or nylon) after which the desired biomolecules are now accessible for further analysis. Other related blotting methods play on the name Southern and include *Northern blotting* (for RNA) and similarly depend on the transfer and detection of material from the gel. An overview of agarose GE in practice is given later in this chapter.

Polyacrylamide gel electrophoresis

Polyacrylamide gel electrophoresis (PAGE) is fundamental to various biomolecular techniques allowing separation of proteins and small fragments of nucleic acids. Arguably the most common use of this technique is in the study of proteins in mixtures and biological samples, separating different proteins into separate bands.

Polyacrylamide gels are prepared using different percentages of acrylamide, which determines the amount of polymerization, and thus gel density, where higher levels of polymerization generate a denser gel structure allowing proteins to separate more easily. A catalyst called *TEMED* controls the polymerization of acrylamide and the cross-linking agent N,N'-methylene-bisacrylamide (see Figure 8.3), and a standard gel for protein separation contains approximately 7.5% polyacrylamide. Gels composed of higher concentrations of acrylamide have relatively smaller pore size, thus allowing analysis of lower molecular weight biomolecules.

After separation of proteins by PAGE, material can be transferred from the gel to a nitrocellulose or poly(vinylidene fluoride) (PVDF) membrane for further

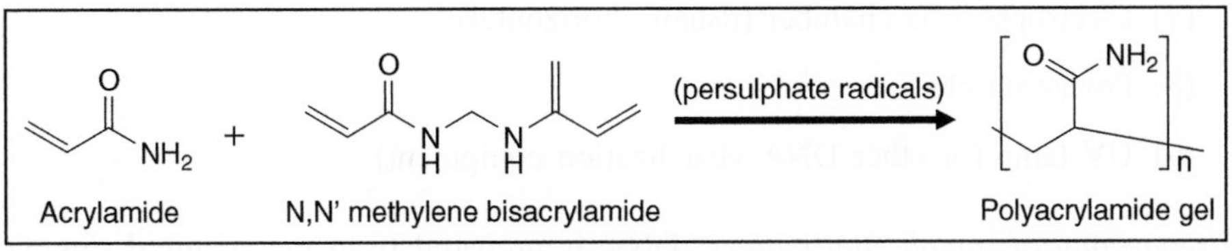

Figure 8.3 **Schematic illustrating the polymerization of acrylamide.**

Western blot analysis. As noted earlier for Northern blotting, Western blotting also plays on the name Southern and is a popular molecular biology method for studying proteins using antibodies and corresponding markers.

8.3 Electrophoresis in practice

As indicated above, electrophoresis represents an important method utilizing an electrical field to move charged particles or molecules with different physical properties through a medium. This section will focus on the set up and use of the basic equipment for analytical gel electrophoresis, which may serve as a precursor (preparative technique) to additional analytical methods including mass spectrometry, PCR and chromatography.

Agarose gel electrophoresis (separation of DNA)

Materials: For this procedure a number of key items are required, which are listed below. It is important to note that, as with many molecular biology techniques, exposure to certain reagents involved in this procedure (such as ethidium bromide) is dangerous, so appropriate personal protective equipment (particularly gloves) is essential.

(1) Sample of DNA to be separated.

(2) DNA ladder–comprising DNA fragments of known size representing molecular weight markers.

(3) Buffer solution.

(4) Agarose gel (and comb).

(5) Ethidium bromide (or other appropriate dye).

(6) Tracking dye (e.g. bromophenol blue) and glycerol.

(7) Electrophoresis chamber (usually horizontal).

(8) Power supply.

(9) UV lamp (or other DNA visualization equipment).

Preparation of sample and buffers: DNA is prepared from the acquired sample using conventional DNA extraction techniques and is mixed with a loading buffer containing bromophenol blue, ready to be loaded into the wells on the agarose gel. Typical buffers used for agarose gel electrophoresis are trisacetate EDTA (TAE) or trisborate EDTA (TBE) used to prepare and run the gel.

Preparation of agarose gel: For many bioanalytical labs the most convenient approach is to purchase high-quality agarose gel available in a precast (set) form, of standard size, which is ready to use. However, some researchers elect to prepare their own gels by the method briefly described as follows: (i) agarose solution prepared in buffer (e.g. 1% agarose solution in TBE); (ii) carefully boil agarose solution (typically using a microwave) and let cool to ~60 °C while stirring; (iii) add minute amount of ethidium bromide to agarose solution and carefully stir well; (vi) pour solution into 'gel rack' and insert comb at one end of gel (~5–10 mm from edge); (v) allow gel to cool and set solid and remove comb to create wells (holes in gel).

Core equipment and methodology: The two primary components required for gel electrophoresis are a buffer-filled electrophoresis chamber (or box) and a separate power supply unit that provides a constant electrical current to the chamber (Figure 8.4). For certain procedures it is also advantageous to have a constant voltage. The agarose gel is transferred into the electrophoresis chamber that contains the buffer solution (e.g. TBE). It is important that the gel is completely submerged in buffer and that the gel is orientated with the wells at the negative electrode. Multiple DNA samples can then be loaded into separate wells and run alongside a *DNA ladder*. The DNA ladder acts as a reference on the basis of the principle that its DNA fragments will migrate through the gel in a defined manner, producing various distinct bands representing rungs of a ladder, hence the name. The DNA samples and DNA ladder both contain the bromophenol blue tracking dye (which is anionic and moves slightly ahead of the DNA) and *glycerol* (to assist the DNA in sinking to the bottom of the well). Typically, the DNA ladder is loaded into the first well (nearest the left-hand edge of the gel) and the other DNA samples are carefully injected into the other wells to run alongside, parallel to the ladder. Once the samples are loaded the electrophoresis chamber is attached to the power supply and electric current applied. DNA is inherently negatively charged because of its phosphate backbone, and hence migrates through the gel from wells at the

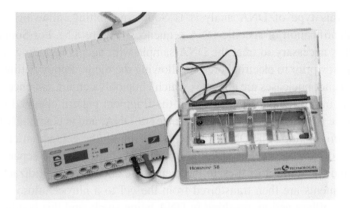

Figure 8.4 **Experimental setup for agarose gel electrophoresis.**

negative electrode (cathode) towards the positive electrode (anode). Conventionally, the smaller the DNA the faster it moves through the gel, and as the tracking dye moves slightly ahead of the DNA, once it approaches the end of the gel the power supply is switched off and migration stops. The gel is then ready for visualization and further analysis.

DNA detection and analysis: Strategically the gel is prepared with ethidium bromide, which is the most popular stain for nucleic acids. Alternatively, ethidium bromide may be incorporated into the buffer, but either way there is sufficient opportunity for it to intercalate between the DNA base pairs during electrophoresis. Dye intercalated with DNA fluoresces when illuminated with (visualized under) ultraviolet (UV) light. It is important to note that there are emerging alternatives to ethidium bromide which are less hazardous and may exceed this dye with regards to sensitivity; for example, Nile blue and cyanine (SYBR) dyes. A *UV light box* is most commonly used, and ethidium bromide dye intercalated with DNA is visualized as a pink colour under UV light. The DNA bands are then visible on the gel, representing different DNA molecules. Each of these bands can be cut out of the gel, from which DNA can be further purified, if required. An important alternative to the use of ethidium bromide staining is *autoradiography*, which requires DNA to be radioactively labelled before it is run on the gel, allowing X-ray identification.

Further DNA analysis: It is also possible to conduct further steps beyond conventional electrophoresis to allow more advanced analyses of the separated DNA bands. Of these steps, blotting is arguably the most versatile tool for further analysis. As outlined earlier, there are various types of blotting, but the most popular related to this type of DNA analysis is Southern blotting, allowing the marking and identification of specific DNA sequences (Figure 8.5). For Southern blot analysis it is necessary to treat the DNA sample with agents including *restriction endonucleases* prior to electrophoresis, following the methodology outlined above. Restriction endonucleases are enzymes which chop only certain sequences of DNA in a sample into fragments. Furthermore, the fragment pattern obtained is usually characteristic for a certain piece (or sequence) of DNA, and is commonly termed a *restriction digest map*. The DNA fragments are run on the gel, and afterwards the gel is treated with an alkaline solution to denature the DNA and separate it into single strands, helping it to stick to the membrane and allow hybridization of the probe. Fragments are then transferred from the gel to a nitrocellulose membrane (sheet) by capillary action, resulting in DNA being stuck, and this bonding can be enhanced by exposure to heat. The membrane is then exposed to a *hybridization probe*. This probe is a specific sequence of DNA that pairs (or hybridizes) with a complementary sequence of DNA on the membrane. As the probe is labelled, it allows detection using autoradiography in the case of a radioactively labelled

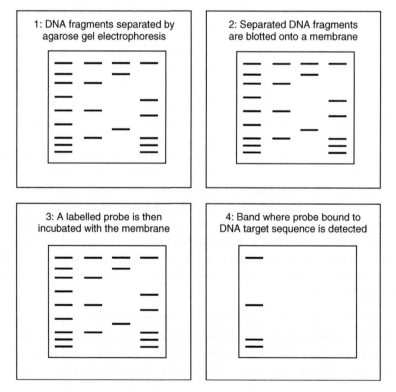

Figure 8.5   Illustration of Southern Blotting technique.

probe, or colouration/fluorescence in the case of a *biotinylated probe*, indicating certain pieces of DNA in the sample.

Polyacrylamide gel electrophoresis (separation of protein)

PAGE may be either non-denaturing or denaturing. The most commonly encountered form is denaturing using substances such as sodium dodecyl sulfate (SDS), and this form of electrophoresis is called *SDS-PAGE*. SDS-PAGE allows the separation of proteins in a gel primarily according to molecular mass. SDS is a hydrophobic anionic detergent which serves to both denature and coat the protein with negative charges prior to electrophoresis. Once the SDS binds the protein, this results in the formation of SDS–protein complexes which have a constant charge to mass ratio–in other words this enables separation on the basis of mass. As such, SDS-PAGE is an important technique to determine the molecular mass of unknown proteins in a sample. SDS–protein complexes move through the gel in a manner where mobility is proportional to molecular mass (i.e. mobility decreases with $\log_{10}$ molecular weight).

Materials: For this procedure a number of key items are required, which are listed below. Again appropriate personal protective equipment should be used, particularly given the toxicity of acrylamide.

(1) Sample of protein to be separated.

(2) Protein ladder–comprising proteins of known molecular weight.

(3) SDS (and commonly a reducing agent such as dithiothreitol or 2-mercaptoethanol).

(4) Buffer solution.

(5) Polyacrylamide gel (and comb).

(6) Electrophoresis chamber (usually vertical).

(7) Power supply.

(8) Protein stain (example Coomassie Brilliant Blue or silver stain).

Preparation of sample and buffers: Protein is prepared from the acquired sample using conventional protein extraction techniques, and treated with SDS to denature the protein, and commonly the protein is also exposed to a *reducing agent*. Reducing agents, such as dithiothreitol or 2-mercaptoethanol, break (or reduce) the disulfide bonds in the protein and, if used, the method is called *reducing SDS-PAGE*. In some cases the approach is to not use a reducing agent, particularly in cases where protein structure is important for further analysis (e.g. enzyme activity in a *zymogram*). The protein is then mixed with so-called Laemmli electrophoresis (running) buffer, comprising Trisglycine (pH 8–9), which also contains a tracking dye ready to be loaded into the wells of the gel.

Preparation of polyacrylamide gel: For many bioanalytical labs the most convenient approach is to purchase high-quality, pre-set polyacrylamide gels of standard size which are ready to use. This is particularly helpful as there are typically two components to a vertical polyacrylamide gel: (i) upper stacking (spacer) gel, and (ii) lower resolving gel (Figure 8.6). The upper *stacking gel* comprises a lower percentage of polyacrylamide (3–5%), has low ionic strength, and near neutral pH. This allows the sample to migrate relatively quickly through the stacking gel and concentrate into thin bands or stacks at the interface with the resolving gel. The lower *resolving gel* has a higher percentage of acrylamide (8–20%), higher ionic strength, alkaline pH, and the protein tends to move relatively slower through

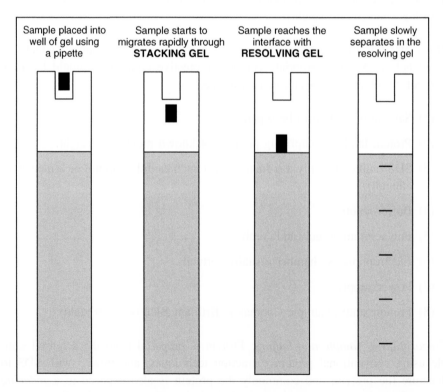

Figure 8.6   Illustration of stacking and resolving gels in polyacrylamide gel electrophoresis.

this component. There are a number of factors that influence the electrophoretic mobility, but principally the smaller pore size of the resolving gel means that the protein mobility is decreased. When preparing the polyacrylamide gel, the resolving gel is poured in between two parallel horizontal plates (clamped together) and allowed to set before the stacking gel is poured on top. A comb is inserted into the top of the stacking gel and when it sets the comb can be removed to create a series of wells, into which samples can be loaded.

Core equipment and methodology: The two primary components required for PAGE are a buffer-filled vertical electrophoresis chamber and a separate power supply unit (Figure 8.7). The gel housed between the two clamped horizontal plates is transferred into the electrophoresis chamber that contains *Laemmli* electrophoresis buffer solution. Multiple samples can be loaded into separate wells and are run alongside a *molecular weight ladder*. The movement of samples and ladder can be monitored by advancement of the tracking dye through the gel. Once the samples are loaded, the electrophoresis chamber is attached to the power supply and electric current applied. The negatively charged proteins migrate through the stacking gel and then onwards through the resolving gel, and once the tracking dye approaches the bottom edge of the gel, the power supply is switched off and migration is halted. The gel is then ready for visualization and further analysis.

Detection and analysis: Once the electrophoretic separation is completed, the gel is treated with a stain, most popularly *Coomassie Brilliant Blue*, to reveal protein bands. In the case of Coomassie Brilliant Blue, this process involves a *stain–destain cycle*, and as proteins can differ in their affinity to take up this dye the detection is really only considered qualitative. There are various other alternative stains which include *Amido black*, *Ponceau red* and *silver nitrate*. Of these, silver staining is widely considered to be one of the most sensitive staining

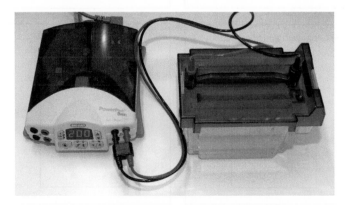

Figure 8.7   **Experimental setup for polyacrylamide gel electrophoresis.**

methods for protein. However, it is possible to use a method where samples are radiolabelled and after gel-derived bands are cut out of the gel, radioactive counts can be measured, giving quantitative information.

Further analysis: After electrophoresis and staining, it is common to undertake further analyses of the separated protein bands, usually by the technique of western blotting (Figure 8.8). Western blotting relies on the transfer of protein from the gel onto nitrocellulose membrane using *electroblotting*, which involves use of a vacuum. The nitrocellulose membrane is then washed with a *blocking solution* (containing an excess of, usually, bovine serum albumin or milk protein, in the presence of a detergent such as Tween 20). This step minimizes non-specific antibody binding prior to incubation with a solution containing a specific antibody against one or more proteins. If the protein is present on the nitrocellulose, the antibody can bind specifically to it, and this is *resolved* by a further incubation with a secondary antibody linked to a so-called *reporter enzyme*, which produces a colorimetric or photometric signal. In most cases, the reported enzyme is *horseradish peroxidase*, which, used in conjunction with a chemiluminescent

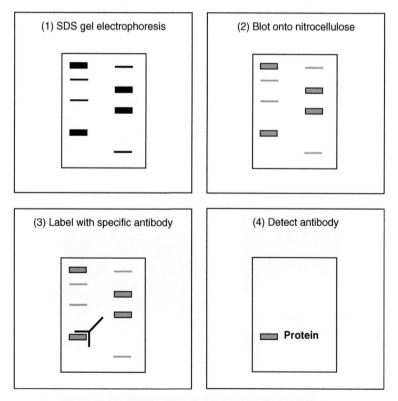

Figure 8.8   **Illustration of western blotting technique.**

agent, produces a luminescence signal proportional to the protein concentration, which can be captured on photographic film.

8.4 Applications of electrophoresis in life and health sciences

The sheer number of ways in which electrophoresis and electrophoretic techniques have been applied means that it is impossible to cover all of the many applications of this powerful bioanalytical tool. This section focuses on a few key applications to give an indication of the versatility of electrophoresis in scientific research and clinical practice.

DNA fingerprinting

Over the years, DNA fingerprinting has been popularized in television dramas, representing an increasingly important forensic tool. This technique has evolved with time, and was originally based on so-called restriction fragment length polymorphism (RFLP) analysis in which DNA in a sample was chopped up using a restriction enzyme. The resulting fragments could then be separated using agarose gel electrophoresis, further analyzed by Southern blotting, and images generated on X-ray film. These images reveal a DNA banding pattern that varies between individuals, thus allowing discrimination among and between species, hence the name DNA fingerprint. Limitations of the RFLP-based technique have drawn into question its reliability and prompted development of other DNA fingerprinting methods incorporating PCR (detailed later), particularly *STR analysis*. STR analysis uses short tandem repeats (STR); that is, sections of DNA that have short repeated sequences, the number of which differs between individuals. The probability of more than one individual having the same profile after STR analysis is exceptionally low, thus this method is considered a very robust approach for DNA fingerprint discrimination.

DNA sequencing

DNA sequencing is popularly termed *Sanger dideoxynucleotide* (or *chain terminator*) sequencing. This method requires double-stranded DNA to be first prepared as a single DNA strand for subsequent sequencing. The single strand is then incubated with DNA polymerase, an enzyme that catalyzes the formation (or synthesis) of a complementary DNA strand from the four bases that make up DNA (i.e. G, A, T, C). The DNA is processed in such a way as to allow labelling (or tagging) of DNA fragments, which can be separated by PAGE and identified by autoradiography, and from this data the DNA sequence can be determined. The

introduction of dye labels for *dye-primer sequencing* allows use of optical identification as opposed to autoradiography–this approach is commonly referred to as
automated DNA sequencing, and was fundamental to the large-scale sequencing
required for the 'Human Genome Project'.

Enzyme activity analysis

Non-denaturing electrophoresis has also found a place in the analysis of enzyme
activity, so-called zymography or activity staining. Zymography is based on SDS-
PAGE, using a polyacrylamide gel incorporating a specific substrate utilized by
the enzyme of interest. This polyacrylamide gel is called a *zymogram*, and the
enzyme–substrate reaction results in a product that can be visualized. The amount
of product is relative to the amount of enzyme present in the gel (i.e. in the
samples), and zymography has been applied to analyze the activities of a diverse
range of enzymes. Another related technique called *reverse zymography* has been
used to examine the activity of enzyme inhibitors.

Clinical protein analysis

Of the many applications of electrophoresis in a clinical setting, arguably one
of the most important is in the analysis of proteins in the blood. This approach
involves electrophoresis of blood serum to separate out the five major serum
protein components (on the basis of size and charge). While *serum protein electrophoresis* is important for in-depth analyses of these protein fractions, *serum
total protein* (or *total protein*) can be measured using another simpler automated
assay. Differences in the levels of total protein or the various protein fractions
determined by electrophoresis are important indicators of disease processes. Low
albumin concentrations may reflect liver disease or acute infection, and high
albumin concentrations may suggest diseases including lymphoma or leukaemia.
Differences in globulin levels are associated with particular disease states, for
example high levels of gamma globulin may indicate the autoimmune disorders
systemic lupus erythematosus or rheumatoid arthritis.

8.5   Advanced electrophoretic separation methodologies
       for genomics and proteomics

The ability to separate nucleic acids and proteins by electrophoresis has enabled
great advances in the emerging fields of genomics and proteomics, which have

much promise for understanding the complex mechanisms underlying normal physiological processes and the pathogenesis of disease. Words ending with 'omics' first emerged in subjects relating to molecular biology, where the suffix '-ome' is generally taken as referring to totality of that topic. As such, the term *genomics* refers to study of the *genome* – that is, the complete genetic makeup of a cell organism. Similarly, the term *proteomics* refers to the study of the *proteome* – that is, a snapshot of the protein composition of a cell/organism at a particular time. The techniques briefly outlined below give indications of how electrophoretic separation technologies have been utilized to advance the fields of genomics and proteomics.

Polymerase chain reaction (PCR)

The introduction of PCR in the 1980s has largely been attributed to the American biochemist Kary Mullis, who shared the Nobel Prize in Chemistry in 1993, recognizing the importance of this landmark method. Mullis' PCR technique realized an idea conceived by Kjell Kleppe and Nobel Prize Laureate Har Gobind Khorana in a paper published in 1971. PCR enables replication of DNA in the laboratory, so-called DNA amplification, producing large quantities of DNA from a small initial sample. After amplification of short DNA fragments by PCR, the PCR products can be separated using agarose gel electrophoresis incorporating a DNA ladder, allowing identification of product fragments. This powerful technique has found numerous uses including gene cloning, detection and diagnosis of diseases, genetic fingerprinting, and paternity testing.

Isoelectric focusing (IEF or electrofocusing)

This method is a type of zone electrophoresis that allows the separation of molecules on the basis of their acidic and basic residues. Separation of molecules occurs along a gel in which a pH gradient has been established. IEF is popularly used to resolve proteins in a sample where a given protein will move through a gel in the presence of an electrical current, stopping at the pH where the net charge of that protein is zero (its *isoelectric point*). When the protein reaches its isoelectric point it is *focused* and visualized as a distinct band. As the isoelectric point is a constant property of a given protein, each protein band can then be identified. Importantly, IEF represents the primary step in the powerful 2D gel electrophoresis method.

2D gel electrophoresis

As the name suggests, 2D gel electrophoresis uses electrical currents to separate biomolecules in a gel in two distinct dimensions–first dimension in a certain direction through the gel and the second dimension in a direction at a 90° angle to the first. The first separation is on the basis of the isoelectric point using IEF, and the second separation is on the basis of size using SDS-PAGE. Use of a second dimension (i.e. 2D) gives enhanced separation on the basis of two properties rather than one. This enhancement should not be understated, as it means that large numbers of proteins from crude cellular extracts can be individually resolved providing an important foundation for proteomic analyses.

Pulsed-field gel electrophoresis (PFGE)

This method is a variant of agarose gel electrophoresis with the primary difference that it utilizes an alternating (as opposed to constant) electrical field. Using an alternating current means that rather than migrating in a straight line, nucleic acid fragments tend to 'slither' through the gel. The principal merit of PFGE is that it allows separation and resolution of much larger DNA fragments than conventional agarose gel electrophoresis. This method is thus suited to examining chromosomal DNA, where it has greatly contributed to understanding of chromosome structure and function. As PFGE allows large scale mapping of chromosomes, it has been utilized in the Human Genome Project and is a gold standard for identification of established and new strains of pathogenic organisms.

Capillary electrophoresis (CE)

As the name suggests, capillary electrophoresis involves electrophoretic separation of biomolecules in the interior of a thin capillary, which may be filled with either a buffer or one of a number of different types of gel. Separation is on the basis of size-to-charge ratio as biomolecules migrate through the capillary and are detected at the capillary outlet. Data arising from CE analysis are displayed as an *electropherogram*, where separated compounds appear as peaks with different retention times (similar to high-performance liquid chromatography (HPLC)), allowing subsequent identification. It is important to note that standard CE separation relies on the fact that the biomolecules possess a charge and that they have inherently different electrophoretic mobilities. CE represents a high-resolution alternative to the use of conventional gel electrophoresis techniques (e.g. SDS-PAGE, IEF), where the capillary is filled with an appropriate gel, hence the term *capillary gel electrophoresis*.

Immunoelectrophoresis (IES)

This technique is a two-step procedure for separation and analysis of protein that exhibits antigenic properties. A sample containing a mixture of proteins is first separated using gel electrophoresis conducted on a small glass plate. An antibody is then loaded onto, and allowed to diffuse through the gel moving across the separated proteins. Interaction between antibody and protein results in the formation of an antigen–antibody complex, which appears as a precipitate (precipitin arc), allowing qualitative analysis. While other techniques that allow quantitative analysis may be preferable, this method has proven valuable in identifying antigenic proteins in blood samples.

Key Points

- Electrophoresis relies on the ability to separate out molecules, in an electrical field, on the basis of them bearing an electric charge.

- Electrophoretic separation occurs because different biomolecules have varying physical characteristics, namely net charge, mass and shape, moving at different rates through a matrix.

- Agarose electrophoresis utilizes agarose gels that have large pores and are used to separate DNA and RNA (200–50 000 base pairs).

- Polyacrylamide gels are most often used to separate small fragments of nucleic acids (up to 500 bp) or proteins (up to 350 000 Da).

- The two primary components required for gel electrophoresis are a buffer-filled electrophoresis chamber (or box) and a separate power supply unit providing constant electrical current.

- Ethidium bromide is a popular stain for nucleic acids, and dye intercalated with DNA during electrophoresis fluoresces when illuminated with UV light.

- Following agarose gel electrophoresis, Southern blotting can be used for more advanced analysis of separated DNA bands, allowing marking and identification of specific DNA sequences.

- Coomassie Brilliant Blue is used following electrophoretic separation to qualitatively stain for protein bands—using radiolabelled samples it is possible to derive quantitative information from bands.

- Following polyacrylamide gel electrophoresis (PAGE), western blotting can be used for more detailed analyses of separated protein bands.

- Electrophoretic separation lies at the heart of polymerase chain reaction (PCR), isoelectric focusing (IEF), 2D gel electrophoresis, pulsed-field electrophoresis, capillary electrophoresis, and immunoelectrophoresis.

9 Applications of mass spectrometry

Mass spectrometry (MS) has emerged as one of the most sensitive methods for detection and evaluation of biomolecules in samples. As the name suggests, it identifies molecules on the basis of their *mass* and *charge*, producing a mass spectrum trace, which is analogous to a molecular fingerprint of the sample. Historically, MS dates back to the late 1890s/early 1900s when the Nobel Prize winning English physicist, Sir Joseph J. Thomson, who discovered the electron, observed deflection of a beam of positively-charged ions which was passed through a combined electrostatic and magnetic field. In these experiments, ions were deflected through small angles, resulting in a series of parabolic curves that were captured on a photographic plate. Each of these parabolic curves corresponded to ions of a particular *mass-to-charge* ratio, with the specific position dictated by the velocity of the ion. This basic setup was later changed, replacing the photographic plate with a metal sheet containing a slit, and by varying the magnetic field a mass spectrum was produced. So, Sir Joseph Thomson could be credited with the invention of MS and the development of the modern mass spectrometer. Since this pioneering work, MS has evolved, producing a complex array of instruments with specialized characteristics. Nevertheless, all MS is based on three key elements: an ion source, analyser of ion beams according to mass–charge ratio, and a detector able to measure or record the currents of the beams. The following sections provide a basic overview of the main principles and key applications of this powerful and versatile bioanalytical technique.

Understanding Bioanalytical Chemistry: Principles and applications   Victor A. Gault and Neville H. McClenaghan
© 2009 John Wiley & Sons, Ltd

Learning Objectives

- To recognize the major types of MS encountered in bioanalytical chemistry.
- To understand the core principles of MS.
- To be able to describe the major types of MS in practice.
- To give examples of MS in bioanalysis.
- To appreciate future uses of MS in life and health sciences.

9.1   Major types of mass spectrometry

Since conception over 100 years ago, MS has become an important analytical and research tool with diverse applications ranging from astronomical study of the solar system to materials analysis and process monitoring in chemical, oil and pharmaceutical industries. Use of MS has led to very many scientific breakthroughs including the discovery of isotopes, accurate determination of atomic mass, and the characterization of biomolecular structure. Indeed, MS is now a fundamental technique employed in pharmacology, toxicology and other biological, environmental and biomedical sciences.

Development of mass spectrometry

Sir Joseph J. Thomson's key research in the late 1890s/early 1900s provided the origin of modern MS. His protégé, Francis W. Aston, developed a mass spectrometer incorporating a new method of *electromagnetic focusing*, which markedly improved resolution and subsequently enabled him to identify most of the naturally occurring isotopes for which he was awarded the 1922 Nobel Prize in Chemistry. Around this time, the American physicist Arthur J. Dempster developed the first electron impact (EI) (spark ionization) source, resulting in an instrument that was considerably more accurate than previous versions. Dempster's work helped establish basic theory and the fundamental design of mass spectrometers which is still used today. In 1946, the American William Stephens presented the concept of a *time-of-flight* (TOF) mass spectrometer, in which ions were separated by differences in their velocities. This TOF-MS method basically determines molecular weight of a biomolecule by accelerating ions toward a detector. The measured time taken for the ion to travel from the source to the detector is very accurately converted to mass, and the greater the mass-to-charge ratio, the slower the ion will travel towards the detector as the TOF-MS accelerates it. In the 1960s the concept of chemical ionization (CI) led the American chemist, Malcolm Dole, to develop electrospray ionization-mass spectrometry (ESI-MS). ESI-MS relies on the evaporation of a fine spray of highly charged droplets in an electrical field, and the resulting ions are drawn into the mass spectrometer. This technique was developed further in the 1980s by the American molecular beam researcher John B. Fenn, who was the first to use electrospray technology to ionize biomolecules. Also, the 1970s saw the introduction of plasma desorption ionization (PDI), where bombarding a sample with a plasma beam resulted in ionization. This means of ionization was superseded by the now popular *matrix-assisted laser desorption/ionization* (MALDI) technique, in which a laser beam (normally nitrogen laser) triggers ionization, and a matrix is used to prevent destruction of the biomolecule(s) by the laser beam. MALDI was developed in the mid 1980s by a team of German scientists led by Franz Hillenkamp and Michael Karas. MALDI offers the advantage that it can ionize biomolecules which by nature are more fragile, quickly losing structure when ionized by other methods. Around this time the Japanese scientist Koichi Tanaka discovered that, using a matrix based on glycerol, direct irradiation of an intense laser pulse on a biomolecule could result in ionization without loss of structure – so-called soft laser desorption (SLD). Tanaka was awarded the Nobel Prize in Chemistry in 2002, which caused some controversy as some would argue that this should have been shared with Hillenkamp and Karas. However, while Tanaka holds the prize for SLD which is not currently used for biomolecules, Hillenkamp and Karas have the satisfaction of knowing that their MALDI technique is the more sensitive method, widely adopted in bioanalytical laboratories worldwide.

The key forms of MS are based upon ESI and MALDI, and these are typically combined with other common analytical separation techniques to give the following powerful bioanalytical tools:

Gas chromatography mass spectrometry (GC-MS): Direct coupling of gas chromatography (GC) and TOF-MS, where the mass spectrometer serves as the detector. Biomolecules are separated by GC and emerge at different times (retention time), passing into the MS which breaks each molecule into ionized fragments, allowing evaluation and identification of separated molecules on the basis of mass-to-charge ratio.

Liquid chromatography mass spectrometry (LC-MS): Direct coupling of liquid chromatography (LC) and MS, where the mass spectrometer serves as the detector. Biomolecules are separated by LC (usually high-pressure liquid chromatography (HPLC)), eluting with different retention times into the mass spectrometer, which can evaluate and identify the biomolecules, again on the basis of mass-to-charge ratio.

Tandem mass spectrometry (MS/MS): Involves three main stages: (i) selection of parent (precursor) ion; (ii) fragmentation of parent ion by so-called collision-induced dissociation (CID); and (iii) analysis of fragmented daughter (product) ions. MS/MS uses mass analysers in series, hence the name tandem mass spectrometry.

Capillary electrophoresis mass spectrometry (CE-MS): Direct coupling of capillary electrophoresis (CE) and MS, where the mass spectrometer serves as the detector. Biomolecules are separated through the narrow capillary of the CE system and then interface with the MS for subsequent analysis.

9.2   Understanding the core principles of mass spectrometry

MS allows identification and study of molecules, relying on the fact that different chemical entities have different masses, charges and other chemical/physical properties. It is extremely important to note that mass spectrometers *DO NOT* actually measure mass, but rather give a record of the mass-to-charge (m/z) ratio. Molecules can be vaporized (forming gas) and also ionized (broken down) into electrically-charged ions, and this represents the first stage of MS. The simple compound salt NaCl comprises sodium ions (Na^+) and chloride ions (Cl^-), each of which have a specific atomic mass and charge. So vaporization and ionization of NaCl produces Na^+ and Cl^- ions which, because of their dissimilar electrical charges, will move differently in an electrical or magnetic field.

   In MS, vaporized ions produced in the first stage enter an acceleration chamber (vacuum) under the influence of a magnetic field. In this second stage, the ions will move through the magnetic field at different velocities, and are deflected in different directions (moving in a curve rather than a straight line; see Figure 9.1) onto a detector. The detector represents the third stage, where the deflection of each ion is recorded, and this data is then used to calculate the mass-to-charge ratio, generating a 'mass spectrum', which allows identification of the biomolecule. As Newton's *Second Law of Motion* dictates that acceleration of a particle is inversely proportional to its mass, lighter ions will deflect further than heavier ones.

   The above refers to a so-called sector instrument, which is a class of MS utilizing a static electric or magnetic sector (or combination of the two) as a mass

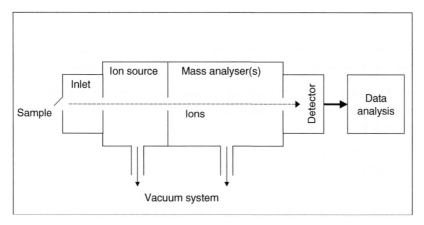

Figure 9.1 Overall mass spectrometer setup and movement of deflected ions.

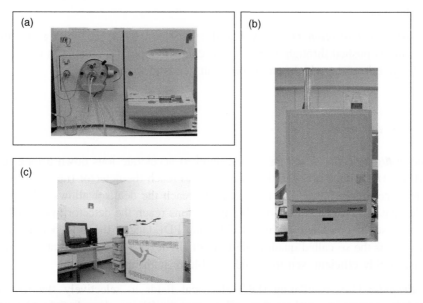

Figure 9.2 Photographs of several mass spectrometers: (a) electrospray ion-trap, (b) matrix-assisted laser desorption/ionization-time-of-flight and (c) nano-electrospray quadrupole-time-of-flight.

analyser. Notably, there are various other types of mass spectrometer which (i) produce different types of ions; and (ii) analyse the ions differently; however, all forms of MS use electrical and magnetic fields to change the paths of ions in some way. Figure 9.2 depicts several major mass spectrometry systems, some of which are discussed briefly below.

Major modes of ionization for MS

Electron impact (EI): On applying a beam of electrons to a biomolecule in a vacuum, the biomolecule loses or gains electrons and is thus ionized.

Chemical ionization (CI): On applying a beam of electrons to a biomolecule in the presence of a large excess of reagent gas (ammonia or methane), the biomolecule gains a proton (H^+) and is thus ionized.

Fast atom bombardment (FAB): On applying a high-energy beam of argon (or zenon) to a biomolecule, it becomes volatile and ionizes.

Plasma desorption ionization (PDI): On applying a plasma beam to a biomolecule, it strips atomic nuclei of their electrons, causing ionization.

Matrix-assisted laser desorption/ionization (MALDI): On applying a high-energy laser beam to a biomolecule (co-crystallized in a matrix), it rapidly turns into a gas and ionizes.

Electrospray ionization (ESI): A liquid (volatile solvent) containing the biomolecule is pushed through a very small, charged capillary, creating a fine spray of charged droplets, which, on evaporation, leads to ionization.

Major MS analysers

Time-of-flight (TOF): Uses an electric field to accelerate ions down a long tube (drift tube) and measures the time they take to reach the detector (i.e. the time of flight) (see Figure 9.3a). The time taken to reach the detector allows calculation of the mass-to-charge ratio. Notably, lighter ions reach the detector first and TOF is often used in conjunction with MALDI and PDI. One of the most popular TOF analysers is the so-called quadrupole-time-of-flight (Q-TOF) configuration, which is fast, highly efficient, sensitive and capable of high resolution.

Quadrupole: Uses oscillating (DC) electrical fields to selectively stabilize (or destabilize) ions passing through a radiofrequency (RF) quadrupole field. Notably, by varying RF it is possible to select for single mass-to-charge ratio values as all other ions will be lost. This type of mass analyser can accommodate higher gas phase pressures than sector analysers and is often interfaced with GC or HPLC.

Ion-trap: This refers to the use of a combination of electric or magnetic fields to capture ions in a vacuum. By varying the precise field conditions, ions are selectively ejected in order of increasing mass-to-charge ratio, and detected

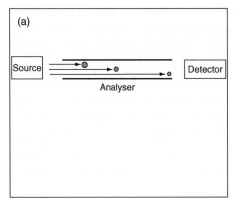
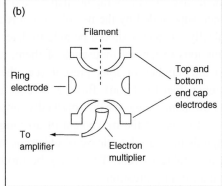

Figure 9.3 **Schematic of (a) time-of-flight and (b) an ion-trap mass analyser. (Adapted from Sheehan, *Physical Biochemistry*, 2000).**

(see Figure 9.3b). There are two main forms of ion-trap mass analysers: (i) quadrupole ion-trap (QIT) analyser and (ii) linear QIT analyser. The QIT analyser is basically the same as the quadrupole mass analyser outlined above, except it traps and ejects ions. The linear QIT analyser is basically the same as the QIT, only it uses a 2D rather than 3D quadrupole field to trap ions.

Fourier transform ion cyclotron resonance (FT-MS or FTICR-MS): This determines the mass-to-charge ratio of ions based on the circular movement (hence *cyclotron*) of ions in an electrostatic magnetic field. Ions are trapped in a *Penning trap* and excited to move in phase in a larger orbit, producing a signal. This signal is mathematically translated by performing a so-called Fourier transform, to produce a mass spectrum. FTICR-MS is a very sensitive technique offering high resolution and precision.

Orbitrap: This determines the mass-to-charge ratio of ions based on their orbit and oscillation around an electrode. The oscillation generates a signal current which is recorded, and the frequencies of the signal current depend on the mass-to-charge ratios of the trapped ions, allowing identification. Fourier transformation allows generation of a mass spectrum. Similar to FTICR-MS, orbitraps (commercially introduced in 2005) are very sensitive and offer high mass accuracy.

MS detectors

The role of the MS detector is to measure the abundance of ions at each mass-to-charge ratio as they are deflected in the MS analyser. The first reported detector made use of photographic paper, and this rather simplistic approach has long been superseded by the use of more advanced and sensitive methods of detection. Ideally a detector should be able to detect any ions in the analyser, that is, from a few ions to tens of thousands of them, and the peak produced should be relative to the number of ions present (i.e. the peak for 50 ions should be 50 times higher than for 1 ion). Basic MS relies on the use of an *electron multiplier*, and as the name suggests this markedly amplifies the signal. The electron multiplier acts to convert the movement of arriving particles (kinetic energy) into electrical signals through the release of electrons from a metal surface (dynode), producing a detectable current. There may be more than one dynode in an MS detector to enhance amplification.

Other types of detector similarly work on the principle of amplification before recording electrical signal. These alternative detectors include: (i) a *scintillator* (*Daly detector*), which incorporates a secondary dynode that emits photons, producing the electrical signal through the use of a photomultiplier tube; (ii) an *array detector*, which allows recording of ions arriving at the same time at different points as opposed to a single point, thus enhancing the efficiency and sensitivity; and (iii) a *microchannel plate (MCP) detector*, where each MCP comprises thousands of individual detection elements, each acting as miniature electron multiplier tubes, enabling the rapid pulse of deflected electrons to be effectively detected over a wide area. MS detectors employed in TOF analysers may also make use of a *time-to-digital converter* (TDC), giving an additional record of time of arrival.

The mass spectrum

As noted earlier, the mass spectrometer gives a record of the mass-to-charge ratios of components of a sample. Graphically, this takes the form of a graphical *plot* or *mass spectrum*, denoting the distribution of molecules (or atoms) in the sample according to their mass-to-charge ratio. The nature of the mass spectrum will depend on how the mass spectrometer handles or modifies the sample components, and while some MS instruments record intact molecules, others break the molecules into fragments (MS/MS). A typical mass spectrum, as illustrated in Figure 9.4, represents a plot of intensity (signal) versus the mass-to-charge (i.e. *m/z*) ratio. Modern MS instruments are under computer control, which both runs the equipment and captures/processes data using specialized software tools. In addition to fully controlling the MS instrumentation, the computer enables the user to acquire the appropriate data and information from MS analysis of the sample.

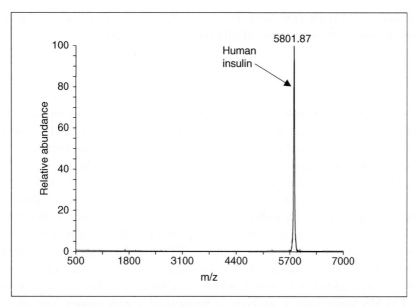

Figure 9.4   **Representative MALDI-TOF mass spectrum of the hormone insulin.**

Other important factors relating to MS analysis

The power and resolution of MS relies on certain inherent features of the MS instrumentation itself. Arguably the two most important factors are *mass resolution* and *mass accuracy*.

Mass resolution: This factor represents the ability of the MS instrument to separate ions with different mass-to-charge ratios. Instruments with high mass resolution are able to provide mass spectra with sharp, clearly defined peaks, allowing discrimination between peaks with similar mass-to-charge ratios that lie very close together on the mass spectrum.

Mass accuracy: As the name implies, this factor represents a comparison between the measured mass (at a set charge) generated by the MS instrument and the theoretical mass (calculated or from tables). The more powerful MS instruments can achieve a high mass accuracy, that is, there is only a small deviation between the measured mass (data from MS instrument) and theoretical mass.

9.3   Major types of mass spectrometry in practice

In practice there are two major and commonly encountered MS tools, namely ESI-MS and matrix-assisted laser desorption/ionization-time-of-flight mass spectrometry (MALDI-TOF-MS). Primarily, these approaches rely on diverse modes of ionization and the use of different mass analysers.

ESI-MS

This widely employed method utilizes electrospray ionization of a sample for subsequent MS analysis, often utilizing an ion-trap mass analyser. ESI-MS generates singly- or multiply-charged ions from the sample solution that are introduced as a fine spray (aerosol) of charged droplets into a strong electric field (see Figure 9.5). Charged ions are produced by one of the following approaches: (i) addition of a proton (hydrogen ion) to the analyte (M) and generation of [M + H] (otherwise written as $[M + H]^+$), (ii) addition of a cation (e.g. sodium ion) to the analyte to generate [M + Na], or (iii) removal of a proton [M–H]. Notably, ESI can also generate multiply-charged ions (e.g. [M + 2H]) and for large macromolecules species [M + 25H] are observed. It is also important to note that electrons are not themselves added or removed, which is common in other forms of ionization. ESI-MS lends itself to analysis of biomolecules with masses typically less than 2000 Da, but has a relatively low mass accuracy. However, this method can be greatly enhanced through interfacing with other separation technologies,

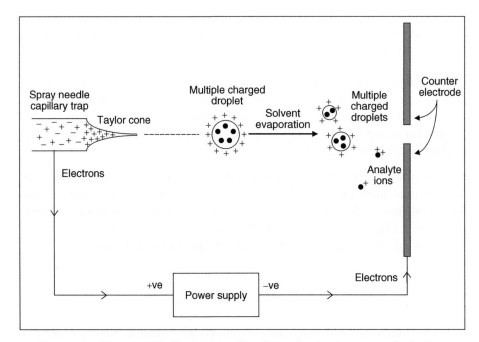

Figure 9.5   **Diagrammatic illustration of ion formation in electrospray ionization.**

particularly liquid chromatography (LC-ESI-MS), and also lends itself to MS/MS for structural analyses.

MALDI-TOF-MS

This sensitive bioanalytical method is based on soft MALDI ionization of a sample, and subsequent analysis using a TOF-MS analyser. Modern MALDI-TOF-MS instruments are highly sensitive and facilitate the rapid and convenient analysis of large numbers (*high-throughput*) of small samples (typically 1 μl). The sample and a matrix (e.g. α-cyano-4-hydroxycinnamic acid) are added to a target plate and allowed to *air-dry* (crystallize) before insertion into the MS instrument. A laser is directed to the crystallized sample causing vaporization and release of ions, which are accelerated in the electric field in a vacuum (*flight tube*). Ions with low mass-to-charge ratios are accelerated to higher velocities and thus reach the detector before ions with higher mass-to-charge ratio (Figure 9.6). As acceleration occurs in a vacuum, velocity is totally dependent on mass, and ions typically reach the detector (*MCP detector*) within 1 ms. Advantageously, MALDI-TOF-MS allows the analysis of large molecules (1000–3000 kDa) without

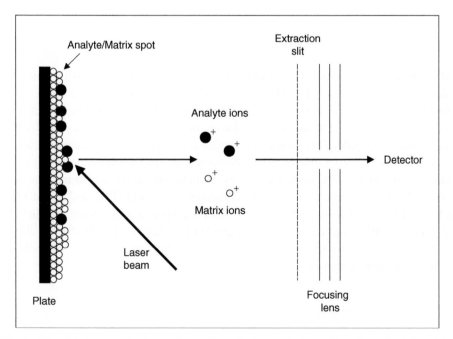

Figure 9.6 Diagrammatic illustration of the mechanism of matrix-assisted laser desorption/ionization.

the need for fragmentation, with an accuracy of 10 ppm. However, MALDI-TOF-MS relies on generation of a single charged ion ([M + H]), and does not lend itself to MS/MS and thus gives little structural information – it is also not compatible with LC-MS. Despite these limitations, MALDI-TOF-MS has rapidly evolved to encompass a wide range of applications in life and health sciences.

9.4   Mass spectrometry: a key tool for bioanalysis in life and health sciences

While straight MS analysis can yield important information in terms of identification, characterization and quantification of biomolecules, it becomes a much more powerful tool with further MS or when combined with other separation technologies. As noted earlier, these approaches include MS/MS, GC-MS, LC-MS and CE-MS. These methods have been extensively exploited in virtually all aspects of bioanalysis, and while fundamentally useful for peptide and protein analysis, these methods have also been used in the analysis of lipids, nucleic acids and a wide range of small molecules and drugs. The range of applications is obviously outside the scope of a book like this, but some indications of the uses of each of these techniques are given below.

MS/MS and peptide identification

MS/MS is often referred to as *Tandem MS*, and is a powerful tool with which to identify proteins. Essentially, MS is performed as usual and this then inputs into another MS analysis, that is, each fragment produced by the first MS is subjected to a second round of fragmentation, and subsequent MS analysis generates an MS/MS fragmentation profile. This profile is complex and requires further computer analysis using specialized software, which compares the fragmentation patterns of the MS/MS spectra with established databases, including those hosted online, accessed via the Internet. The very large *online databases* (e.g. SEQUEST and MASCOT) allow direct comparisons between the observed spectra and known (predicted) spectra, thus allowing sequencing and peptide identification. For MS/MS analysis of larger proteins the fragmentation is aided by the use of specific proteolytic enzymes including trypsin (see Figure 9.7).

GC-MS and hormone analysis

GC-MS has proven a particularly useful tool to measure hormone levels in biological fluids. A key example of the use of this approach has been in the screening

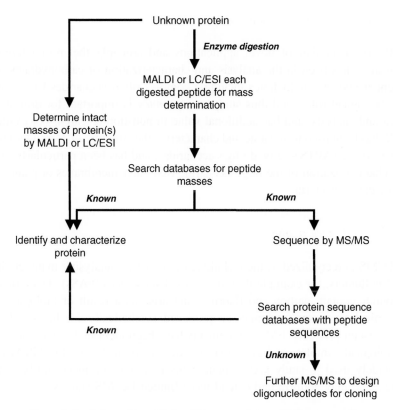

Figure 9.7   **Schematic outlining MS approach to peptide sequencing.**

and detection of performance-enhancing steroids in urine. The first use of GC-MS in this context was in 1976 at the Montreal Olympic Games, in testing athletes for banned steroid drugs. Testament to the success of this approach, GC coupled to high resolution MS was used to screen every specimen taken during the Atlanta Olympic Games in 1996.

LC-MS and clinical therapeutic drug screening

LC-MS has emerged as an important tool for the identification and quantification of therapeutic drugs and/or metabolites in a range of biological samples (e.g. blood, urine, cerebrospinal fluid and hair) of animals or humans. Examples of agents that have been analysed by LC-MS include amphetamines, opiates, benzodiazepines, non-steroidal anti-inflammatory drugs (NSAIDs), and poisons such as nicotine or pesticides. As such, LC-MS is a useful technology for clinical or forensic toxicology.

CE-MS and oligosaccharide analysis

CE-MS has a number of novel applications and arguably the most interesting among these has been in the analysis and characterization of carbohydrates from different species. Carbohydrates including complex oligosaccharides have a number of biological roles, and thus study of *glycomics* is important for drug development and analysis, and has additional value in nutrition research. For example, CE-MS has been used to separate and characterize the structure of 1-aminpyrene-3, 6,8-trisulfonate (APTS)-derived oligosaccharides, and has been particularly useful in the characterization of the oligosaccharide-rich outer membranes of pathogenic Gram-negative bacteria.

LC-MS/MS and trace analysis

LC-MS/MS is recognized as the technique of choice for analysing many environmental pollutants, for example fluorinated alkyl substances (FASs). These organic compounds contain carbon and fluorine, and arise as a result of biological and geochemical processes, and may be produced synthetically for heat-, oil- and water-resistant products. FAS contaminants have been identified in environmental and biological samples, and can bioaccumulate to toxic levels. LC-MS/MS has been widely used to study FAS contaminants in water, sediment and biological samples, and may confirm and extend more limited LC-MS studies.

Other separation technologies used with MS

In theory, any form of separation can occur before MS analysis but in practice the most important of these, apart from those outlined above, is the combined use of 2D gel electrophoresis with MS. This approach has been used to characterize the complex protein mixtures encountered when examining samples of tissues, cells and subcellular fractions. This combination has come into its own in large-scale automated proteomics, including that of plants (e.g. German Plant Genome Project). For this procedure proteins separated on 2D gels (protein spots) are excised and digested prior to MS analysis, and spectra analysed using specialized software.

9.5   Mass spectrometry: future perspectives

As outlined above, MS in its various forms can be utilized as a powerful research and diagnostic tool with emerging clinical applications. As with many other bioanalytical technologies, MS is constantly evolving, with developments in new technologies and applications paralleling the rise of the 'omics' and in particular genomics, proteomics, lipidomics and metabolomics. Fundamental methods such as MALDI-TOF and ESI-MS are now being superseded by more sophisticated instrumentation and software, including the use of biochips and powerful data mining approaches to analysis. For example, FTICR-MS, mentioned earlier, is becoming increasingly recognized and utilized for the characterization of *in vitro* cellular systems, currently providing exceptionally high quality data for analysis of biological material. Furthermore, coupling advanced instruments with other sensitive techniques, including high-performance micro-capillary column separation, provides innovative means of improving detection, identification, quantification (e.g. sensitivity and dynamic range), and throughput. Paralleling these improvements in instrumentation and approach are advances in method development and, in particular, matrices used and sample preparation. In this regard, anticipated future incorporation of *nanomaterials* in sample pre-treatment before MS will revolutionize the analysis of biological samples.

It is also important to note that the scope of MS is not confined to small molecules, and intact biomolecules and their complexes can also be analysed, providing new information including stoichiometry and binding affinity of proteins. Such developments are accelerated by the advent of so-called ambient MS techniques, allowing compounds ranging from biopolymers and small drugs to endogenous biomolecules to be analysed in crude samples. Additional developments relate to the size of the MS instruments, and there is a progressive move towards miniaturization, with more compact devices such as a Microflex or MALDI microMX system providing a more versatile platform for the 'omics'. This has stimulated interest in portable or handheld MS devices allowing *in situ* analysis; particularly useful for on-site forensic testing at crime scenes.

Key Points

- Mass spectrometry (MS) identifies molecules on the basis of their mass-to-charge (m/z) ratio, producing a mass spectrum, analogous to a molecular fingerprint of the sample.

- Mass spectrometers comprise three key elements: ion source, mass analyser (according to mass-to-charge ratio) and detector (measuring and recording output).

- Vaporized ions enter an acceleration chamber under the influence of a magnetic field, move at different velocities, and deflection measures are used to calculate m/z ratio.

- Key forms of MS such as ESI and MALDI can be combined with other analytical separation techniques including GC-MS, LC-MS, MS/MS and CE-MS.

- Modes of MS ionization are EI, CI, FAB, PDI, MALDI and ESI; and major MS analysers include time-of-flight (TOF); quadrupole; ion-trap (QIT); Fourier transform ion cyclotron resonance (FTICR); and orbitrap.

- A typical mass spectrum represents a plot of intensity (signal) versus m/z ratio, and modern MS instruments capture and process data using specialized software tools.

- ESI-MS generates singly- or multiply-charged ions (e.g. through addition [M + H] or removal [M–H] of a proton), allowing analysis of biomolecules less than 2000 Da.

- MALDI-MS utilizes a matrix on a target plate, which is vaporized using a laser (producing single charged ions [M + H]) before acceleration and detection.

- MALDI-MS allows analysis of larger biomolecules without fragmentation with high sensitivity/accuracy, enabling rapid high-throughput analysis of small sample volumes.

- Tandem MS (MS/MS) has proven a particularly powerful tool, comparing fragmentation patterns of MS/MS spectra with established databases, enabling protein identification.

10 Immunochemical techniques and biological tracers

There are a range of immunochemical techniques and biological tracers that are important tools for any bioanalytical laboratory. As the name implies, immuno-chemical techniques (immunoassays) are based on key chemical reactions associated with components of the immune system. These methods are routinely used in biochemistry laboratories to identify and measure amounts of a particular substance in body fluids, particularly blood or urine. Immunoassays are principally based on the reaction of an antibody with an antigen, where a specific antibody is used to identify and measure a given antigen or vice versa. A number of different tracers can be used to follow biological pathways and monitor distribution of molecules in the environment and complex samples. Historically, the Swedish chemist Svante August Arrhenius, in his famous text published in 1907, coined the term *immunochemistry*, and somewhat ironically he died in 1927 from an attack of acute intestinal catarrh, likely of immunological origin. While the 1940s saw the emergence of diagnostic immunochemical tests, it was not until the 1950s that Yalow and Berson developed the now famous and widely used radioimmunoassay (RIA) technique. Yalow was later recognized with the Nobel Prize in Physiology or Medicine *'for creating the Yalow–Berson method to measure minute amounts of peptide hormones using antibodies'*, some five years after her death; and if she or Berson had patented the immunoassay, they would have made a fortune! The following sections outline the principles and key applications of a range of popular immunochemical techniques, and the use of biological tracers in life and health sciences.

Understanding Bioanalytical Chemistry: Principles and applications   Victor A. Gault and Neville H. McClenaghan
© 2009 John Wiley & Sons, Ltd

Learning Objectives

- To recognize the importance of antibody–antigen interactions in immuno-chemical measurements.

- To illustrate the analytical applications of biological tracers.

- To describe and explain the principles and applications of radioimmunoas-say (RIA).

- To be able to convey knowledge of the principles and applications of enzyme-linked immunosorbent assays (ELISA).

- To appreciate immunohistochemistry (IHC) as an important diagnostic tool.

10.1   Antibodies: the keys to immunochemical measurements

The reaction between an *antibody* (or antibodies) and an *antigen* lies at the heart of immunochemistry. An antibody is a biomolecule of the *immune system*, which targets, binds to, and neutralizes/eliminates a foreign object, which is the anti-gen. The immune system represents the body's defence mechanism, creating and maintaining barriers that prevent infectious agents from wreaking havoc on body systems. There are two main arms to the immune system, so-called *innate* and *adaptive*. The adaptive immune system relies on the antibody–antigen interaction, which provides the key immunochemical elements and the basis of immunoassays.

What is an antibody?

Antibodies are large Y-shaped glycoproteins and members of the so-called immunoglobulin superfamily. There are five main types of antibody, namely immunoglobulin A, D, E, G and M (IgA, IgD, IgE, IgG and IgM). Each of these differs in their size, charge, amino-acid composition, carbohydrate content and thus biological properties. However, they all share a common purpose, which is to bind specifically to a given antigen. This binding occurs at the tips of the Y-shaped antibody molecule, each of which acts as a '*lock*' to which the antigen '*key*' attaches. This *lock-and-key* analogy is useful in understanding specificity, as each lock has only one key. Antibodies are produced by B-lymphocytes from the humoral immune system, and *humoral immunity* is the term used to refer to the production of antibodies and all the processes accompanying it. As such, humoral immunity is mediated by antibodies that bind antigens on the surface of *foreign bodies* such as microbes (viruses, bacteria), marking them for destruction. The word humoral was derived on the basis of involvement of substances in body fluids, historically termed *humours*.

Analytical immunochemical techniques require the production of antibodies, which is primarily achieved by using animal models. *Polyclonal antibodies* (antisera) are produced by administration of a given antigen to normal laboratory animals, commonly through injection into guinea pigs or rabbits. Often *booster* injections are given to optimize antibody yield, but after a suitable time period blood samples are taken, and after clotting the liquid phase (blood serum) is removed, from which antibodies are prepared and purified. *Monoclonal antibodies* are prepared using cultured plasma clonal cells that provide a rich source of antibodies of known and uniform specificity.

What is an antigen?

Simply put, antigens are minute substances that evoke an immune response and are generally classified as *immunogens*, *tolerogens* and *allergens*, each of which stimulates antibody production. While antigens are usually proteins or polysaccharides, they can be any type of molecule, including *haptens* (small molecules) covalently bound to *carrier proteins* (which facilitate transport in blood and into tissues or cells).

How do antibodies and antigens interact?

Understanding the interaction between antibodies and antigens relies on basic knowledge of the structure of an antibody (an immunoglobulin molecule). Antibodies are comprised of immunoglobulin (Ig) units with *monomeric* (one Ig unit), *dimeric* (two Ig units; e.g. IgA), *tetrameric* (four Ig units; e.g. fish IgM), or *pentameric* (five Ig units; e.g. mammalian IgM) forms (see Figure 10.1). For simplicity, we will focus on the monomer and its binding properties.

Monomer

- Y-shaped molecule

- Consists of four polypeptide chains (i.e. two heavy and two light chains)

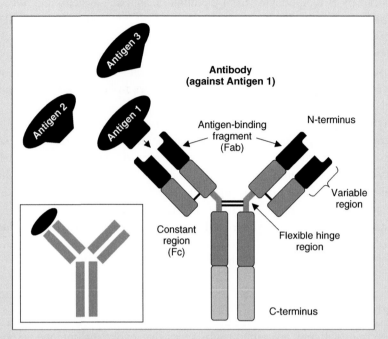

Figure 10.1 Schematic of an antibody molecule and its interaction with an antigen.

- The heavy chains or light chains are connected by disulfide bonds

- There are five types of heavy chain (γ, δ, α, μ and ε)

- There are two types of light chain in mammals (λ and κ)

- Stem of Y contains the Fc (fragment, crystallizable) region, which binds to various cell receptors and complement proteins, regulating antibody function

- Each end of Y is called the Fab (fragment, antigen binding) region

- Fc and Fab fragments can be generated for experimental/analytical use

Antigens bind to the *Fab* region of the antibody, and depending on the particular antibody there may be one binding interaction (*monovalent*) or several simultaneous binding interactions (*multivalent*). The strength of binding interaction of a single antibody with a single antigen is called the *affinity*, whereas the term used to describe more than one binding interaction is called *avidity*. The latter is not the true affinity and may be several orders of magnitude greater than the affinity, so these terms are not interchangeable.

Antibody–antigen interaction at the heart of immunoassay

Antibodies have three major properties, which provide the basis and characteristics of immunoassays:

(1) They bind a wide spectrum of chemicals, biomolecules, cells and viruses (*range*).

(2) They inherently bind specific substances (*specificity*).

(3) They form a strong interaction with their target (*strength of binding*).

These properties are important when considering the use of antibodies in immunoassay techniques, and if they did not show such properties they would not be useful diagnostically and would not survive the processing steps in the assay procedure. Further enhancement of immunoassays comes from the use of identical antibodies, so-called monoclonal antibodies. Generally, monoclonal antibodies only bind at one target site of a given molecule, thus increasing specificity and accuracy, particularly in complex biological samples. However, it is important to realize that immunoassays can be designed to measure either antigen or antibody. For example, an immunoassay to measure insulin in a sample of blood will employ a chosen antibody that binds insulin. So in this case the immunoassay is designed to measure an antigen (insulin). When the body is trying to fight an infection it will produce high levels of antibodies to destroy a specific infectious agent (*pathogen*). In this case the immunoassay employs a chosen pathogen (antigen) to measure blood levels of antibody that bind that particular pathogen.

Measurement of antigens or antibodies in immunoassays

As described above, antigens or antibodies can be detected using *specific immunoassays*, but in order to quantify the levels of antigen or antibody in a sample (*unknown*) the immunoassay procedure must include a reference (*known*). This reference must also effectively and specifically bind to the antigen or antibody under investigation, and is called a '*standard*'. Typically, the result obtained from the binding of different known amounts (concentrations) of this standard enables the preparation of a so-called standard curve. This standard curve is plotted as a graph over a range of known measured values, and thus the measured value of the unknown can be elucidated. Depending on the assay, the known measured values (*signal*) plotted as a standard curve may either increase (curve upwards) or decrease (curve downwards). However, in either case the immunoassay is critically dependent on the ability to determine from the standard curve the signal generated by the unknown sample (i.e. the unknown value is within the range of the plotted standard curve).

Dynamics of antibody–antigen interactions

Immunoassay relies on the binding of antigen to antibody, and this reaction may take seconds to hours to reach a *steady state* or *equilibrium*, influenced by various factors including ionic strength, pH and temperature. Assays are often designed with these factors in mind to optimize formation of antibody–antigen complexes. The reaction between antibody and antigen is simplistically described below, following the *Law of Mass Action*:

$$[Ag] + [Ab] \quad \underset{K_d}{\overset{K_a}{\longleftrightarrow}} \quad [Ag\text{–}Ab]$$

From this equation, a given concentration of antigen ($[Ag]$) interacts with a given concentration of antibody ($[Ab]$) to form an antigen–antibody complex ($[Ag\text{–}Ab]$). This is a dynamic reaction which can occur in both directions, where K_a is the so-called *association rate* constant working in opposition to K_d, which is the so-called *dissociation rate* constant.

Antibody–antigen reactions and assay development

There are several key types of antibody–antigen reaction that are important to understand, which are as follows:

Agglutination tests: This represents one of the oldest measures of antigen–antibody reaction. This method gives both qualitative and quantitative data, relying on the interaction of antibodies with a suspension of particulate antigen. The formation of antibody–antigen complexes is visible and seen as *clumping*, termed *agglutination*. This method is commonly used to identify microbes or in haemagglutination tests for blood group identification.

Precipitin tests: As the name implies, precipitin tests rely on the fact that when the appropriate ratio of antibody and antigen are mixed together, immune complexes of antibodies and soluble antigens come out of solution, settling to form a visible *precipitate*. The antibody–antigen precipitate formation can be plotted as a curve, and interaction is maximal at the top of the curve (termed the *zone of equivalence*) shown in Figure 10.2. This technique can be used to quantify the antibody content of a solution.

Liquid-phase and solid-phase reactions: Antibody or antigen reactions can occur where both components are in the same liquid phase (*in solution*), or one component is in the solid phase and the other is in the liquid phase (*solid–liquid interface*). The former liquid phase assays were the first type used but have since largely been superseded by various forms of solid-phase immunoassay. Solid-phase assays

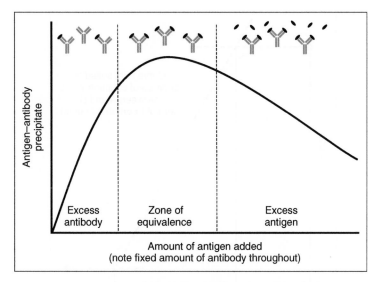

Figure 10.2   **Diagrammatic illustration of a precipitin curve.**

rely on reactions between antibody and antigen at a solid–liquid interface. When a solution containing an antigen (liquid phase) is applied to antibody bound to a solid phase or surface (e.g. multi-well test plate), the antibody–antigen reaction takes place at the solid–liquid interface. Alternatively, of course, the solution may contain the antibody and the solid phase or surface the antigen.

Assays inherently require the ability to detect unknown levels of antibody or antigen in a biological sample and they may be *competitive* or *noncompetitive*, and can also be *heterogeneous* or *homogeneous*.

Competitive immunoassays: This class may also be referred to as *limited reagent* assays. These assays have the following major components: (i) *specific antibody*; (ii) *sample containing unknown amount of antigen*; and (iii) *labelled antigen*. The assay relies on the operator mixing (i), (ii) and (iii) together either *simultaneously* or *sequentially* and allowing the opportunity for (ii) and (iii) to compete for binding to (i). In parallel, the operator prepares a reaction which uses (i) and (iii), but this time using (iv) *known standard antigen*. In this way the operator can observe the competition between (iii) and (iv) for binding to (i), thus establishing a standard curve for data analysis and determination of (ii). A typical standard curve, which is essentially a binding dose–response curve, is given in Figure 10.3.

Noncompetitive immunoassays: This class comprises the following subtypes: *excess reagent*, *two-site* and *sandwich* assays. A typical noncompetitive assay has the following major components: (i) *primary or capture antibody*; (ii) *sample*

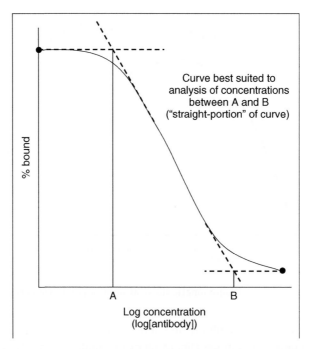

Figure 10.3   Typical dose–response curve for an immunoassay.

containing unknown amount of antigen; and (iii) *secondary or labelled antibody*. The assay typically relies on the operator coating a multi-well test plate (solid phase) with (i), and after this process (ii) is added, allowing the antibody–antigen reaction to occur. After a washing step, (iii) is then added to each well and incubated for a given time to allow reaction (conjugation) with (ii), which is attached to (i). It is important to note that (iii) will bind a different region of (ii) than does (i). These are typically kit-based assays, which provide all reagents including (iv) *a known standard antigen*. From (iv) a standard curve is prepared allowing the concentration of (ii) to be determined. Notably, the data arising from the noncompetitive assay is different than that of the competitive assay in that, noncompetitive assays give measures of labelled antibody bound which are directly proportional to the concentration of antigen, as when antigen is not present the labelled antibody will not bind.

Heterogeneous or homogeneous immunoassays: Certain assays require an additional method to remove *unbound label* (also termed *free label*) leaving *bound label*, and for a given assay these labelled components are either antibody or antigen, but not both. These assays are called heterogeneous assays, and common methods of separation include adsorption, precipitation, or commonly the use of a solid phase. In contrast, homogeneous assays do not require the removal of unbound from bound label to make a measurement, and thus are more convenient to perform. One of the most popular heterogeneous immunoassays is ELISA, which is covered in more detail later. Two important homogeneous immunoassays are *enzyme multiplied immunoassay technique* (EMIT) and *cloned enzyme donor immunoassay* (CEDIA). EMIT has been used to develop a wide range of clinically-relevant assays for drugs, hormones and metabolites, and complex CEDIA assays are based on a bacterial *Escherichia coli*-generated enzyme called *β-galactosidase*. For CEDIA assays, two distinct parts of the β-galactosidase protein are engineered, and measurements are dependent on the ability of these to assemble when mixed together, a process modulated by the amount of antigen or analyte, generating colour.

Major Labelled Immunoassays

Radioimmunoassay (RIA): These assays rely on the use of radiolabels. Typically, antigen is labelled using a radioactive isotope, for example iodine-125 (^{125}I), and radioactivity (radiation emitted from tube) is measured using an appropriate counter (in this case a γ radiation counter). This major technique and some applications are covered in more detail later.

Enzyme immunoassay (EIA): These assays exploit the catalytic properties of enzymes. Typically, antibodies labelled with an enzyme are used, for example horseradish peroxidase. The enzyme, which is bound and remains after washing, is able to convert added substrate to generate a coloured product that can be measured. The major enzyme immunoassay (EIA) is ELISA, which is covered in more detail later.

Fluoroimmunoassay: These assays rely on the use of fluorescence-emitting labels. Typically, antibodies labelled with a *fluorophore* are used, for example, fluorescein isothiocyanate. After washing, measured fluorescence relates to the amount of antigen. Another useful variant is called the homogeneous *fluorescent polarization immunoassay* that is commonly used to measure drugs and analytes.

Chemiluminescence immunoassay: These assays rely on the use of chemiluminescent labels, which may be labelled antigens or antibodies. For example, the chemiluminescent label isoluminol is oxidized in the presence of hydrogen peroxide and a catalyst, producing long-lived light emission that is measured, thus allowing determination of the unknown antigen or antibody concentration in a sample. Another useful variant is electrochemiluminescence immunoassay that is commonly used in biomolecular detection, and in particular the measurement of native and recombinant peptides and proteins.

Protein microarrays: These emerging methods are based on simultaneous multianalyte immunoassays, normally using enzyme or fluorophore labels. As such, these provide the opportunity to conduct diverse multiple tests on single test plates or *glass or plastic chips*. This technology is based on arrays of thousands of minute (micrometer) amounts or dots of antibody or antigen attached to a solid phase. As the name suggests, this is only useful for protein–protein interactions, and detection is by chemiluminescence or fluorescence using a scanning instrument that can read samples very quickly.

Simplified immunoassay: This class of immunoassay has been developed for so-called point-of-care testing, in devices such as home pregnancy-test kits. As such, simplified immunoassays integrate the power of molecular immunology with advanced material and processing sciences to allow convenient, rapid, accurate and sensitive testing by untrained users. A major example of the widespread use of simplified immunoassays is in the measurement of human chorionic gonadotropin (hCG; protein hormone) in urine samples using a labelled anti-hCG antibody in a one-step procedure.

10.2   Analytical applications of biological tracers

There are a number of tracers that have been used to help understand chemical reactions and interactions. Historically, development of modern tracer methods began with the pioneering work of the Hungarian physical chemist, George Charles de Hevesy, in the early 1900s. De Hevesy's work focused on the use of radioactive tracers to study chemical processes, for which he was awarded the Nobel Prize in Chemistry in 1943. Radioactive tracers, also known as *radioactive labels*, are based on the use of a given *radioisotope*. However, it is important to note that there are also *isotopic tracers* (or *isotopic labels*). *Isotopes* are forms of a chemical element with different atomic mass, which have nuclei with the same atomic number (i.e. number of protons) but different numbers of neutrons. Examples include: ^{3}H, ^{14}C, ^{32}P, ^{35}S and ^{125}I, which are radioactive forms of stable elements ^{1}H, ^{12}C, ^{31}P, ^{32}S and ^{127}I. The word isotope comes from Greek, meaning 'at the same place', a useful way to remember that all isotopes of an element are in the same place in the Periodic Table of elements. While not quite correct, often the words isotope and *nuclide* are used interchangeably. Thus, radioisotopes may be termed *radionuclides*. The later refers to an atom with an unstable nucleus that undergoes radioactive decay, and these may be naturally occurring or artificially produced.

Radioactive decay

Historically, the discovery of radioactivity dates back to 1896 when the French scientist Henri Becquerel believed that the *afterglow* observed in *cathode ray tubes* might be associated with *phosphorescence*, later realizing that this phenomenon was instead due to radiation. At first, this radiation was assumed to be similar to X-rays, but further research by Becquerel and a number of other notable scientists (including Marie Curie and Ernest Rutherford) revealed that the nature of this radiation was more complex. Subsequently, it emerged that there were three principal forms of radioactivity that result from different types of radioactive (nuclear) decay.

During radioactive decay an *unstable atomic nucleus* emits radiation in the form of particular particles or electromagnetic waves. This process results in a parallel loss of energy as so-called parent nuclide(s) transform into daughter nuclide(s). The principal types of radioactive decay are *alpha* (α), *beta* (β) and *gamma* (γ), as described further in Table 10.1; the SI unit of radioactive decay is the *Becquerel* (Bq), where one Bq is one *decay* (or *transformation/disintegration*) per second.

The Bq is a minute measure of radioactivity and any sizeable amount of radioactive material will contain very many atoms and thus emit considerable amounts (TBq or GBq) of radiation. Another popular unit of decay is the *curie*, a non-SI unit (historically calculated from the disintegrations of radium) which is equivalent to 37×10^{10} Bq. Importantly, radioactivity decays exponentially, where a population of atoms in a sample will have a characteristic *half-life* ($t_{1/2}$). The half-life is the key parameter when considering radioactivity and associated safety of radioisotopes, where $t_{1/2}$ represents the time taken for the radioactivity to fall to a half the recorded level, as illustrated in Figure 10.4. Half-lives and associated properties of common radioactive isotopes are given in Table 10.2.

Measurement of radioactivity

Radioactivity can be measured by a variety of methods producing values which relate to Bq (i.e. one disintegration per second). Due to inherent limitations of certain instruments which cannot measure each and every disintegration in a sample (i.e. efficiency <100%), often the measure of decay is recorded as *radioactive counts* per unit time—typically 'counts per minute' (or cpm). In most cases these instruments will additionally correct for background

Table 10.1   Properties of major types of radiation

Type of radiation	Penetration range in air (m)	Shielding material
Alpha (α)	0.025–0.080	Not necessary
Beta (β)	0.15–16	Plastic (i.e. Perspex)
Gamma (γ)	1.3–13 (intensity half)	Lead

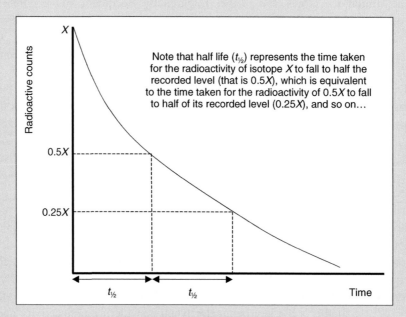

Note that half life ($t_{1/2}$) represents the time taken for the radioactivity of isotope X to fall to half the recorded level (that is $0.5X$), which is equivalent to the time taken for the radioactivity of $0.5X$ to fall to half of its recorded level ($0.25X$), and so on…

Figure 10.4   Illustration of radioactive decay and half-life.

Table 10.2   Radiation properties and uses of key isotopes in biology

Isotope	Radiation	Half-life	Major uses
^{125}I	γ	59.9 d	Labelling biomolecules such as peptides, proteins and nucleotides
^{14}C	β	5715 yr	Labelling organic molecules
^{32}P	β	14.3 d	Labelling proteins and nucleotides
^{35}S	β	87.2 d	Labelling proteins and nucleotides
^{3}H	β	12.3 yr	Labelling organic molecules

radiation (noise), presenting count data as *disintegrations* per minute (dpm). The two most popular methods for measuring and monitoring radioactivity are outlined below.

Geiger–Müller (GM) tube: The sensing element of a so-called *Geiger counter* (named after inventor Hans Geiger), which detects single particles of ionizing radiation (gaseous ionization detector), emitting characteristic audible clicks (via an audio amplifier) that intensify as the level of radioactivity increases (Figure 10.5a). The number of pulses per second represents the intensity of the radiation and is principally confined to the detection of γ radiation.

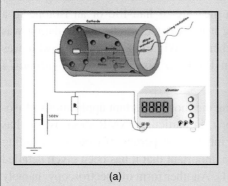

(a)   (b)

Figure 10.5 Diagram of a Geiger–Müller tube (a) and photograph of a liquid scintillation counter (b).

Scintillation counter: The sensor, the so-called scintillator, contains a transparent crystal that fluoresces when hit by ionizing radiation, thus a scintillation counter measures ionizing radiation. Light emitted from the crystal is measured by a sensitive photomultiplier tube which is attached to an electronic amplifier in order to count the amplitude of signals produced by the photomultiplier. Liquid scintillation counters are a very efficient and practical way to measure and quantify β radiation (see Figure 10.5b).

Properties of isotopes define their application

Isotopic labelling and substitution: Less commonly encountered isotopes of an element (*trace isotopes*) provide a useful way to follow chemical reactions as they usually behave in a similar way to stable elements. Referred to as *isotopic labelling* or *isotopic tracing*, this allows, for example, a glucose molecule (containing six carbon atoms) to be labelled with one or more isotopic carbon atoms (^{13}C, ^{14}C). As the glucose is metabolized, the metabolic derivatives become enriched with the isotopic carbon label and are detected using advanced technologies such as mass spectrometry and nuclear magnetic resonance (NMR) spectroscopy. Importantly, incorporation of isotopes into molecules can alter reaction mechanisms including the rate of a reaction–this is termed *isotopic substitution* and forms the basis of the kinetic isotope effect.

Radiometric dating: As radioisotopes have a given half-life they can be used to calculate or date a sample. Referred to as *radiometric dating* (including *carbon dating*), it is possible on the basis of a measure of half-life of an isotope in a sample to determine its age or how long it has existed. Radiocarbon dating was discovered in 1949 by the American chemist, Willard Frank Libby, who was awarded the Nobel Prize in Chemistry in 1960 *'for his method to use carbon-14 for age determination'*.

Exploitation of nuclear properties: Among the most important applications of isotopes is in advanced forms of spectroscopy. In particular, NMR spectroscopy has emerged as a powerful tool utilizing the inherent properties of isotopes including ^{1}H, ^{15}N, ^{13}C and ^{31}P. This technique is so important that it has been given detailed coverage in a separate chapter (Chapter 11). Another form of spectroscopy, namely *Mössbauer (Mößszligbauer) spectroscopy*, relies on absorption of γ radiation and nuclear transitions of given isotopes, most commonly ^{57}Fe. Radionuclides are also fundamental to particle accelerators (e.g. *cyclotron*), *nuclear reactors* and *power stations*, and development of nuclear weapons using large quantities of isotopes including uranium and plutonium. Other uses include *autoradiography* and other important methods of nuclear medicine such as *positron emission tomography (PET)*. Radionuclides also have important applications in food preservation, agri-culture, animal husbandry, industrial welding, mining, archaeology, palaeontology, and analysis of environmental pollution.

10.3   Principles and applications of radioimmunoassay (RIA)

RIA is one of the primary techniques which has historically been important for clinical knowledge and testing for hormones. Indeed, RIA was first developed by Yalow and Berson to measure the circulating levels of the important hormone insulin. However, since this early description of the technique, very many similar assays have appeared for hundreds of substances differing markedly in chemical structure and biological activity.

As noted earlier, RIA typically depends on the use of radiolabelled antigens, and in the case of the first report in the scientific literature the antigen in question was insulin. This classic radioimmunoassay for insulin will be used as the fundamental example of how this technique operates in practice.

Principles of radioimmunoassay

RIA relies on the basic principles of antigen–antibody complex formation outlined earlier in the chapter.

Reagent preparation: All RIA procedures require pre-assay preparation of (i) *specific antibody (Ab)*, (ii) *labelled antigen (Ag*)*, (iii) *standard antigen (Ag)*. **Ab** is raised in animals and should be of high specificity, to allow the sensitivity of the method to approach that required to quantify compounds in the picogram (10^{-12} g) to nanogram (10^{-9} g) level. Ag* is prepared by one of a number of techniques including the *iodogen method*, *chloramine T procedure*, *lactoperoxidase*, and an indirect method using *Bolton–Hunter reagent* (especially nucleic acids). Ag is prepared by isolation and purification of the antigen in question. In this case, insulin standard is prepared from homogenization of the pancreas (which contains the insulin-producing cells) and acid–ethanol extraction of insulin, which is subsequently purified by high-performance liquid chromatography (HPLC), freeze dried and weighed.

RIA procedure: Traditionally, a **specific antibody** (**Ab**) is incubated in the presence of **labelled antigen** (Ag*) to allow **Ab** to be saturated, that is, form a complex with Ag* (**Ab**–Ag* complex). This Ag* is the labelled equivalent of the **antigen** (Ag) to be measured in the sample, and after Ag is added there is a competitive interaction where some Ag* may be displaced from the **Ab**–Ag* complex and replaced by Ag (i.e formation of **Ab**–Ag). The amount of Ag* that is displaced will reflect the amount of Ag in the sample. As noted earlier, a standard curve is constructed where known amounts of standard antigen (Ag) are incubated with the same amount of **Ab**, where Ag displaces Ag* from the **Ab**–Ag* complex, thus allowing parallels to be drawn between the displacements which form **Ab**–Ag (sample) and **Ab**–Ag (standard). This process is illustrated in Figure 10.6.

Separation for measurement: The basis of measurement depends on the ability to separate the unbound or '*free*' Ag* from the **Ab**–Ag* complex, which itself relies on differences in properties between the two components while maintaining (that is not disrupting) the **Ab**–Ag* complex (Figure 10.6). There are a number of reagents used for separation including *charcoal*, *hydroxyapatite*, *ammonium sulfate* and *polyethylene glycol*, but the method that will be considered here relies on the use of *dextran-coated activated charcoal* (see Figure 10.7). With inherent

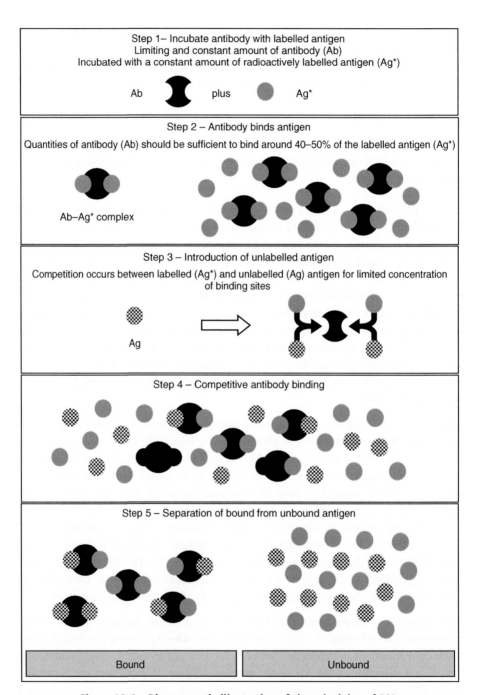

Figure 10.6   Diagrammatic illustration of the principles of RIA.

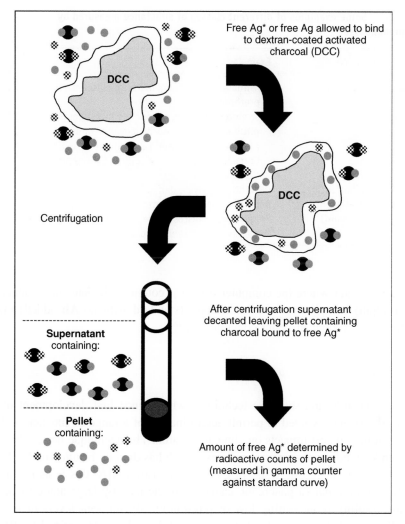

Free Ag* or free Ag allowed to bind
to dextran-coated activated
charcoal (DCC)

DCC

DCC

Centrifugation

After centrifugation supernatant
decanted leaving pellet containing
charcoal bound to free Ag*

Supernatant
containing:

Pellet
containing:

Amount of free Ag* determined by
radioactive counts of pellet
(measured in gamma counter
against standard curve)

Figure 10.7   Illustration of dextran-coated charcoal method for separating bound and unbound antigen.

ability to adsorb small molecules, dextran-coated activated charcoal is able to bind free Ag*, thus enabling separation from the **Ab**–Ag* complex. This is achieved by centrifugation, where the charcoal which binds free Ag* sediments to the bottom of the test tube, while **Ab**–Ag* complex remains in the supernatant and can be discarded. This leaves a pellet which represents the charcoal bound to free Ag*, and the amount of free Ag* can be determined by the radioactive count of the pellet. Measurement depends on the relative amounts of radioactivity in the sample versus the standard curve. Radioactive counts are typically measured

Table 10.3   Some examples of different classes of substance measured by radioimmunoassay

Class of substance	Examples
Haematological agents	Erythropoietin, fibrinogen/fibrin, plasminogen/plasmin, prothrombin
Hormones	Gastrointestinal (e.g. vasoactive intestinal peptide; VIP), pituitary (e.g. adrenocorticotropic hormone; ACTH), pancreatic (e.g. insulin), parathyroid (e.g. parathyroid hormone), reproductive (e.g. hCG), thyroid (e.g. thyroxine)
Nucleic acids/nucleotides	DNA, RNA, cytosine derivatives
Pharmacological agents	Amphetamines, barbituates, morphine, nicotine
Steroids and vitamins	Progesterone, Testosterone, vitamin B_{12}, vitamin D
Viruses	Hepatitis, *Herpes simplex*, polio, rabies

using a γ counter where the computer-driven γ counter calculates the 'counts per minute (cpm) bound' from the 'total cpm' (standard with no **Ab** added) minus the 'free cpm' (count of pellet containing Ag*).

Applications of radioimmunoassay

RIA is a versatile and sensitive technique, which since first development in the 1950s to1960s has evolved to permit determination of a range of substances, having one of the biggest impacts on clinical research. This technique has contributed to the understanding of disease processes, which has directly resulted in the development of new diagnostic tests. In the case of insulin RIA, this is as useful for detecting *insulinoma* (a pancreatic cancer characterized by large amounts of circulating insulin) as measuring low or physiologically relevant levels of insulin. An indication of the range of substances which have been determined by RIA is given in Table 10.3 below.

10.4   Principles and applications of enzyme-linked immunosorbent assay (ELISA)

The ability to measure minute amounts of substances such as insulin represented a key breakthrough in the field of *endocrinology*, and development of RIA was key to the later development of ELISAs. Indeed, ELISA has superseded many RIA measures with its inherent advantages associated with the avoidance of the use of radioactivity and ability to conduct automated analyses with colorimetric signals.

The most popular form of this technique is solid-phase heterogeneous ELISA, and the inherent use of passive adsorption facilitates flexibility in assay design. ELISA is generally classified as: *direct*, *indirect*, *sandwich* (*double-antibody*) or *competitive*. Principles of each of these formats are briefly discussed below.

Direct ELISA

There are two forms of direct ELISA, namely labelled-antibody and labelled-antigen. Both follow a similar protocol, but as the former is arguably the most widespread it is described below (Figure 10.8).

General stepwise procedure

Step 1: Antigen adsorption to solid-phase support

Step 2: Washing

Step 3: Addition of enzyme-linked antibody

Step 4: Incubation

Step 5: Washing

Step 6: Addition of specific substrate

Step 7: Incubation and colour development

Step 8: Colorimetric measurement

Step 9: Data analysis

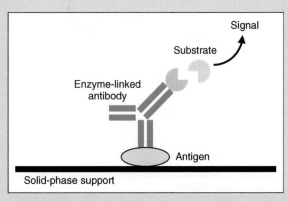

Figure 10.8 Diagrammatic overview of direct ELISA.

Indirect ELISA

This method is extensively used for detection of specific antibodies from serum samples. One major advantage of this form of ELISA is that the enzyme-linked secondary antibody can detect more than one primary antibody, as bound antigen and not the secondary antibody determines specificity (Figure 10.9).

General stepwise procedure

Step 1: Antigen adsorption to solid-phase support

Step 2: Washing

Step 3: Addition of primary antibody directed against antigen bound to solid-phase

Step 4: Incubation

Step 5: Washing

Step 6: Addition of secondary enzyme-linked antibody directed against primary antibody

Step 7: Incubation

Step 8: Washing

Step 9: Addition of specific substrate

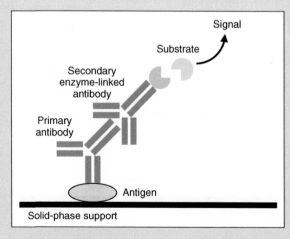

Figure 10.9   Diagrammatic overview of indirect ELISA.

Step 10: Incubation and colour development

Step 11:Colorimetric measurement

Step 12: Data analysis

Sandwich (double-antibody) ELISA

This method is normally used to detect specific antigens, and is particularly useful for virus detection. Importantly, this technique exists in two forms (direct and indirect), both of which rely on the adsorption of antibody as opposed to antigen. The technique involves three major components, namely a capture antibody, antigen and secondary enzyme-linked antibody (Figure 10.10).

General stepwise procedure

Step 1: Capture antibody adsorption to solid-phase support

Step 2: Washing

Step 3: Addition of antigen

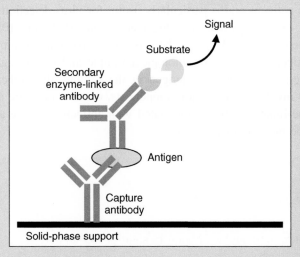

Figure 10.10   Diagrammatic overview of sandwich (double-antibody) ELISA.

Step 4: Incubation

Step 5: Washing

Step 6: Addition of secondary enzyme-linked antibody (detection antibody) directed against antigen bound to capture antibody

Step 7: Incubation

Step 8: Washing

Step 9: Addition of specific substrate

Step 10: Incubation and colour development

Step 11: Colorimetric measurement

Step 12: Data analysis

Competitive ELISA

While competitive assays often imply that two agents are trying to bind to a single, third agent, proper competition occurs when both agents are added simultaneously. Inhibition or blocking assays rely on the sequential addition of the two competing agents (although in practice one agent may be added and incubated before the second 'competitor' is added). This method is normally used to detect specific antigens and is particularly useful for virus detection. Importantly, this technique exists in two forms (direct and indirect). Figure 10.11 illustrates a direct competitive ELISA where there is an inverse relationship between the amount of antigen present in the biological sample and colour signal measured (that is, the less the amount of antigen in the sample, the higher the signal). The technique involves three major components, namely a capture antibody, secondary enzyme-linked antibody conjugated with antigen, and free antigen (agent in sample to be measured by assay).

General stepwise procedure

Step 1: Capture antibody adsorption to solid-phase support

Step 2: Washing

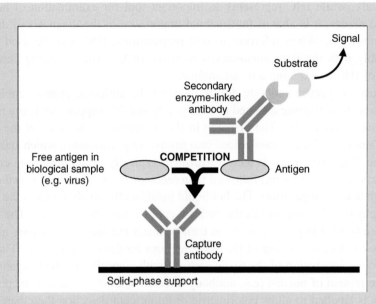

Figure 10.11   Diagrammatic overview of competitive ELISA.

Step 3: Simultaneous addition of secondary enzyme-linked antibody conjugated with antigen and free antigen (unknown sample, e.g. virus)

Step 4: Incubation

Step 5: Washing

Step 6: Addition of specific substrate

Step 7: Incubation and colour development

Step 8: Colorimetric measurement

Step 9: Data analysis

10.5   Immunohistochemistry: an important diagnostic tool

Immunohistochemistry (IHC) is a generic term that describes techniques used to determine the presence of one or more antigenic proteins in cells and/or tissues. Samples suitable for immunohistochemical analysis include *smears* and *swabs*, cells either cultured or in suspension, and thin sections of body tissues.

Importantly, while IHC represents an important tool for examination of healthy cells and tissues, it is also of fundamental importance in the diagnosis of disease (*histopathology*). When referring to cell preparations, IHC may be used interchangeably with the term *immunocytochemistry* (ICC). The following describes the use of IHC to examine a tissue sample.

IHC uses a specific antibody directed against the antigenic protein, applied to cells or a section of tissue attached to an optically suitable support, such as a microscope slide or cover slip. The first step in IHC is removal (*excision*) of the tissue to be examined, which is then placed into *fixative* (e.g. formalin), which stabilizes the tissue. The fixed tissue is then *dehydrated*, typically by immersing in ethanol (of increasing concentration) followed by *toluene* or *xylene*, before it is *embedded* (e.g. with hot liquid paraffin). The hot liquid paraffin effectively replaces the water in the dehydrated tissue and as the paraffin cools it hardens into *wax*. The tissue sample embedded in paraffin wax can then be finely cut into thin *sections* using a *microtome* before *de-waxing* of the tissue sections for further processing. The next step involves incubation of the tissue sections with a specific antibody against the antigenic protein of interest (e.g. antibody directed against insulin). If the primary antibody is labelled to allow colorimetric or fluorescent detection then there is no need for further washing steps and incubation with secondary antibody. However,

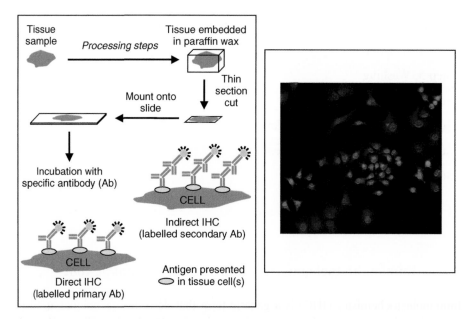

Figure 10.12 **Steps involved in immunohistochemistry and a photomicrograph of pancreatic β cells stained for human insulin with a fluorescent label and visualized by fluorescent microscopy.**

use of a secondary enzyme-labelled or fluorescent-labelled antibody may be necessary, but in either case the presence of the antigenic protein in a tissue section is usually visualized by either light or fluorescent microscopy (see Figure 10.12).

IHC is an excellent method for protein localization and detection in a tissue sample; however, its usefulness relies greatly on primary antibody specificity and, as such, some care may be required in interpretation. An invaluable tool in diagnostic hospital and research laboratories, IHC plays a fundamental role in clinical diagnostics and has proven particularly useful in the identification, localization and profiling/typing of tumours (e.g. carcinomas). In the research setting, IHC has been widely employed in examination of normal and abnormal protein expression in the brain (e.g. Alzheimer's disease).

Key Points

- Antibodies are large Y-shaped glycoproteins (immunoglobulins) produced by the immune system, denoted IgA, IgD, IgE, IgG or IgM.

- Antigens are minute substances which stimulate the immune system to produce antibodies which bind to them for neutralization/elimination.

- The adaptive immune system relies on the antibody–antigen interaction, which provides the key immunochemical elements and the basis of immunoassays.

- Antigens bind to the Fab region of the antibody, and strength of binding interaction of a single antibody with a single antigen is called the affinity.

- Antibodies have three major properties, which provide the basis and characteristics of immunoassays, namely: range; specificity; and affinity (or strength of binding).

- Immunoassays detect and quantify unknown levels of antibody or antigen and may be competitive or noncompetitive and heterogeneous or homogeneous.

- Major labelled immunoassays include: RIA, EIA, fluoroimmunoassay, chemiluminescence immunoassay, protein microarrays, and simplified immunoassay.

- RIA relies on the use of radiolabelled tracers and the ability to separate the unbound or 'free' radiolabelled antigen (or antibody) from the antigen–antibody complex.

- ELISA is a colorimetric immunoassay which may generally be classified as: direct, indirect, sandwich (double-antibody) or competitive.

- Immunohistochemistry is a generic term that describes techniques used to determine the presence of one or more antigenic proteins in cell and tissue samples.

11 Bioanalysis by magnetic resonance technologies: NMR and MRI

The principle uses of magnetic resonance phenomena are in biomedical research and clinical diagnosis. There are two major methods that exploit the magnetic properties of biomolecules in quantitative and qualitative bioanalysis, namely nuclear magnetic resonance (NMR) spectroscopy and magnetic resonance imaging (MRI). NMR relies on being able to differentiate the intrinsic magnetic properties of an atom's nucleus, its so-called *intrinsic magnetic moment*. MRI evolved from NMR and was originally called *nuclear magnetic resonance imaging* (NMRI) and thus incorporates the same basic principles discussed in the following sections. Historically, the development of NMR came from the independent research of the American physicist, Edward Mills Purcell, and Swiss-born physicist, Felix Bloch. Purcell's research at Harvard University resulted in the discovery in the mid 1940s of NMR, while Bloch's work focused on nuclear induction and NMR, which form the basis of MRI technology, one of the most important medical advances of the last century. Both these scientists shared the 1952 Nobel Prize in Physics for *'their development of new methods for nuclear magnetic precision measurements'*. It should be stated from the outset that technologies based on magnetic resonance phenomena are particularly complex and intellectually challenging, and the authors hope that the following sections provide a concise, informative and accessible outline of the principles and key applications of NMR and MRI technologies for the life and health sciences.

Understanding Bioanalytical Chemistry: Principles and applications   Victor A. Gault and Neville H. McClenaghan
© 2009 John Wiley & Sons, Ltd

Learning Objectives

- To be aware of the differences between NMR, MRI and related technologies.

- To describe and explain the principles of NMR.

- To illustrate the established and emerging applications of NMR.

- To convey knowledge of the principles and uses of MRI.

- To demonstrate applications of MRI in diagnosis and research.

11.1   Nuclear magnetic resonance (NMR) and magnetic resonance imaging (MRI) technologies: key tools for the life and health sciences

Nuclear magnetic resonance is a physical phenomenon that lies at the heart of both NMR spectroscopy and MRI. However, these two technologies differ in their applications, and while NMR spectroscopy is widely adopted for identification and characterization of biomolecules, MRI is used for non-invasive visualization (imaging) of the inside of living organisms. In order to understand these technologies it is important to consider their historical development and draw comparisons with other related techniques.

History of Nuclear Magnetic Resonance

The discovery of NMR dates back to the mid 1940s and is based upon fundamental research by Bloch and Purcell. Both scientists spent the years of the Second World War working on science pre-dating NMR technologies. While Bloch was researching atomic energy at Los Alamos National Laboratory, Purcell was involved in the development of *microwave radar* at the Massachusetts Institute of Technology. Both scientists undertook further research at Harvard University, with interests and expertise that contributed to RADAR. The term RADAR is an acronym for **ra**dio **d**etection **a**nd **r**anging, coined in the early 1940s, where radio waves detect/determine direction, distance and speed of objects, and allow these to be mapped accordingly. The study of RADAR technology, based on the production, detection and absorption of *radiofrequency* (RF) energy, likely contributed to the understanding of NMR. These studies revealed that *magnetic nuclei* of elements such as ^{1}H and ^{31}P absorb RF energy from magnetic fields. As these magnetic nuclei absorb RF energy they are excited from a lower energy state (E_1) to a higher energy state (E_2), and the energy required to achieve this is termed the *resonance energy*. In physics, *resonance* is described as the tendency of an object to *oscillate* with *high amplitude* when *excited* using energy of a certain frequency. So when a magnetic nucleus absorbs energy it is in resonance. It is important to note that atoms in biomolecules are not identical and thus different atoms will resonate at different frequencies in the same magnetic field. Given the above, knowledge of NMR has paralleled the growing understanding of electromagnetic phenomena and technologies such as infrared (IR) spectroscopy. Modifications to original NMR approaches utilizing *continuous-wave (CW) spectroscopy* included the introduction of *faster pulsed* NMR techniques in the early 1950s, allowing enhancement of this technology. Advances included the introduction of multidimensional *Fourier transform nuclear magnetic resonance spectroscopy* (FT-NMR) following the pioneering work of Swiss chemist Richard R Ernst, who was later awarded the Nobel Prize in Chemistry in 1991. Later, in 2002, Kurt Wüthrich shared the Nobel Prize in Chemistry for his leadership in the development of multidimensional NMR spectroscopy for the study of proteins, and NMR technologies have greatly impacted on the life and health sciences.

Development of magnetic resonance imaging

As noted earlier, MRI technologies emerged as an offshoot from NMR, and one of the first potential commercial applications in this context is documented in a 1974 US patent filing by American physician Raymond V Damadian. This followed Damadian's paper, published in the prestigious journal *Science* in 1971, where he noted different responses of tumour tissue and normal tissue to NMR. While some of these initial methods were flawed, this was nonetheless a significant contribution to the principal that NMR and related scanning technologies could be used in a diagnostic context. However, Damadian's research built upon earlier work by the American physicist Herman Carr, who created a one-dimensional (1D) magnetic resonance image. These innovations prompted American chemist Paul Lauterbur and British physicist Sir Peter Mansfield to develop Carr's technique to produce the first two-dimensional (2D) and three-dimensional (3D) MRI images. Although NMR is the primary scientific principle underlying MRI, both Lauterbur and Mansfield were awarded the Nobel Prize in Physiology or Medicine in

2003 as it was not until later that NMR was exploited to produce body images. However, their being awarded the Nobel Prize caused some controversy, and it has been argued that Carr and/or Damadian should also have shared this most prestigious honour. This is more acute considering that it was actually Damadian who produced the first full MRI scan of the human body and was also regarded as being the person who discovered NMR tissue relaxation differences that made NMR scanning feasible. Interestingly, the Lemelson-MIT program bestowed its Lifetime Achievement Award (2001) on Damadian as 'the man who invented the MRI scanner'. Damadian has also won large patent cases against MRI manufacturers for infringing his MRI patents, even though Damadian's methodology is not used in modern MRI imaging and diagnostics, adding further to the controversial history of MRI.

Alternative techniques to NMR and MRI

Electron spin resonance (ESR): An alternative to NMR is the technique of ESR that relies on the measurement of *electron spins* as opposed to nuclear magnetic spins, which is the basis of NMR. The basic principles on which ESR is based are similar to those of NMR, but ESR technique detects unpaired electrons in a molecule. As molecules with unpaired electrons are referred to as *paramagnetic*, ESR is often alternatively termed *electron paramagnetic resonance* (EPR). While ESR lends itself to analysis of free radicals or transition metals, that is, molecules that characteristically have unpaired electrons, most stable molecules do not have the *unpaired spin* required for ESR analysis. Also, energy differences in ESR are measured in the microwave region of the electromagnetic spectrum as opposed to the radiowave region associated with NMR, and ESR data is generated using an ESR spectrometer. While the bioanalytical scientist more commonly encounters NMR, examples of the diverse uses of ESR include characterization of *reaction pathways*, for example the *mitochondrial electron transport chain* in cells and *photosynthetic pathways* in plants. Medical applications include insertion of so-called *spin labels* into biomolecules to render them paramagnetic and thus allowing tagging and tracking via ESR.

Computed tomography (CT): An alternative approach to imaging is CT, otherwise known as *computer axial tomography* (CAT/CT scan) or *body section roentgenography*. While MRI uses radiofrequency signals and analysis, CT scanning involves the use of *ionization radiation* (that is X-rays) for image analysis. CT is thus a powerful medical imaging tool using tomography (imaging by sections) and *digital geometric analysis* to produce 3D images from a large number of acquired 2D X-ray images. As CT uses X-rays it is particularly useful for imaging dense tissues such as bone, whereas MRI is most commonly used to examine soft (that is non-calcified) tissues. While both CT and MRI have good *spatial resolution*, MRI has a much enhanced *contrast resolution*, discussed in more detail below.

11.2 Principles of NMR and the importance of this biomolecular analytical technique

NMR is a physical phenomenon based on magnetic properties of atoms and their nuclei. Atoms contain three major particles, namely electrons, neutrons and protons. Atomic nuclei contain neutrons and protons, and those with odd numbers of neutrons or protons possess a *magnetic moment* (μ) arising from *nuclear magnetic spin*. As nuclei spin they create a magnetic field with a particular *spin moment* and *intrinsic angular momentum*. These magnetic nuclei will exhibit certain characteristics in an *external magnetic field* (B_0), where they *orient*/align *parallel* to (with) or *anti-parallel* to (against) the external field (see Figure 11.1). These

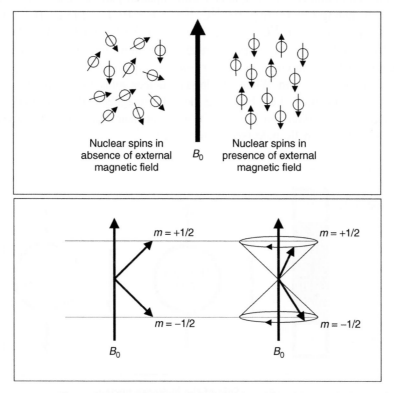

Figure 11.1 Alignment of magnetic nuclei when placed in an external electric field.

orientations/alignments of individual nuclei are referred to as *spin states*, which have different equal and opposite *magnetic quantum numbers* (m_I), where the m_I value is either + (when parallel) or − (when anti-parallel). Notably, the parallel orientation/alignment is favoured energetically.

The collective magnetic spin of nuclei results in a defined overall nuclear spin quantum number, denoted as I. Notably, elements (e.g. carbon or hydrogen) can have isotopes (e.g. ^{12}C, ^{13}C or ^{1}H, ^{2}H) which may have different nuclear spin quantum numbers (e.g. ^{12}C, $I = 0$; ^{13}C, $I = 1/2$; ^{1}H, $I = 1/2$; ^{2}H, $I = 1$). Also, nuclei possessing paired particles (i.e. have even atomic number/mass) have no net spin and thus $I = 0$. For example, the commonly observed nuclei ^{12}C (six neutrons and six protons) and ^{16}O (eight neutrons and eight protons) have no angular momentum or magnetic moment, and as $I = 0$ they do not show up when analysed by NMR (i.e. cannot produce an NMR spectrum).

In contrast, usually a nucleus with odd numbers of protons and neutrons has an integral non-zero nuclear spin quantum number (i.e. $I > 0$). This arises because the total number of unpaired neutrons or protons is even, and each contributes 1/2 to the quantum number. As such, nuclei with unpaired particles can have different I values, including $I = 1/2$ (e.g. ^{1}H, ^{13}C, ^{15}N, ^{19}F, ^{31}P), $I = 1$ (e.g. ^{2}H, ^{14}N), $I = 3/2$ (e.g. ^{11}B, ^{23}Na, ^{35}Cl, ^{37}Cl), $I = 5/2$ (e.g. ^{17}O, ^{27}Al), and $I = 3$ (^{10}B).

Nuclei with $I = 1/2$ have spin angular momentum and associated magnetic moment (μ) which can either be $+\mu$ or $-\mu$ (i.e. equal and opposite), depending on the direction of the spin. When a hydrogen nucleus spins, the positively charged proton forms a magnetic field behaving like a bar magnet with a north and south pole, and the spin direction (i.e. direction of nuclear rotation) determines the direction of the magnetic field (see Figure 11.2).

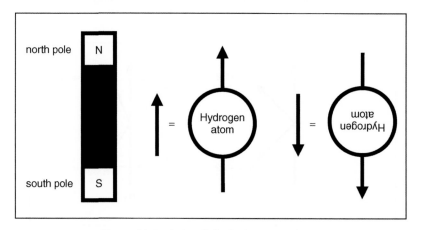

Figure 11.2   Spin of the hydrogen nucleus.

Resonance

It is the possession of a magnetic moment which is very sensitive to its surroundings and can be manipulated that allows nuclei to be analysed by NMR spectroscopy. In NMR, the magnetic moment is altered by the application of electromagnetic radiation of a defined frequency. At this point it is important to note that resonance is a term used to describe the tendency of a given system to oscillate with high amplitude when excited by energy at a particular frequency. In other words, when magnetic nuclei are impacted by electromagnetic radiation from a high frequency transmitter there is a change in their movement.

When placed in a magnetic field, a magnetic nucleus will adopt one of a small number of orientations of different energy. In the case of a hydrogen atom the nucleus (a proton) has two permitted orientations, where the magnetic moment points in the same direction as the field (parallel) or the opposite direction (anti-parallel). As the nuclei absorb this electromagnetic energy they are promoted from a *lower energy state* (E_1) to a *higher energy state* (E_2), the energy required for this to occur is called the *resonance energy* $(\Delta E = E_2 - E_1)$. The resonance energy (ΔE) depends on the strength of interaction between the nucleus and field, that is, the size of the nuclear magnetic moment and magnetic field strength. In order to measure the *energy gap*, ΔE, electromagnetic radiation of particular frequency (ν) is applied to the hydrogen nucleus, causing it to '*flip*' from E_1 to E_2. Under these conditions, resonance of a specific nucleus (in this case hydrogen nucleus) gives rise to a nuclear magnetic resonance spectrum, fundamental to NMR analysis.

Every nuclide (e.g. ^1H, ^2H, ^{13}C) has a characteristic magnetic moment in a magnetic field of particular strength ΔE, and thus the resonance frequency is determined by the nuclide. Also, the resonance frequency depends, in part, on the chemical environment of the nucleus in a molecule, the so-called *chemical shift*. A simple example of chemical shift, reported by J. T. Arnold and co-workers in 1951, relates to three different kinds of H atom in liquid ethanol (chemical structure CH_3CH_2OH), which give rise to three different resonance frequencies, identified on a ^1H NMR spectrum as three separate peaks (and three separate areas) in the ratio 1:2:3 ($OH:CH_2:CH_3$), reflecting the number of protons of each type, hinting at how NMR can be used to probe individual atoms in molecules. Chemical shift is, however, not the only source of information encoded in an NMR spectrum, and magnetic interactions between nuclei (*spin–spin couplings*) can give rise to extra NMR lines. In fact, when recorded on an appropriate spectrophotometer, the spectrum of liquid ethanol can have not three but eight (or more) distinct peaks. That is, under stringent conditions, the OH singlet can split into a triplet and each line of the CH_2 quartet can become a doublet, giving rise to extra NMR lines.

Relaxation

After a short time, nuclei in the higher energy state (E_2) will return to the lower energy state (E_1) by a process called *relaxation*. When considering relaxation, the time taken is a critical factor, where T_1 time (the so-called spin–lattice relaxation time) is the average time taken for an individual nucleus to return to its equilibrium state (when it can be probed again) and T_2 time (the so-called spin–spin relaxation time) is the lifetime of the observed NMR signal. In the NMR spectrum, T_2 time determines the width of the NMR signal, where nuclei with large T_2 times give rise to sharp signals and those with shorter T_2 times have broader signals. Relaxation effects have structural and dynamic applications, where T_1 measurements are useful for dynamic (motional) studies, while the so-called nuclear Overhauser effect can give structural information (such as internuclear separations). Together these approaches can give lots of information about the molecules being studied.

Principles of NMR spectroscopy

NMR spectroscopy exploits the magnetic properties of nuclei to give important qualitative and quantitative information on biological samples. There are various different forms of this important technique, but in each case NMR spectroscopy relies on fundamental NMR theory discussed earlier.

There are three essential requirements for NMR recordings: (i) an intense and static (stable) magnetic field with defined field strength; (ii) a source of RF radiation (high frequency transmitter) to irradiate and excite nuclei in the sample; and (iii) a method or device to detect the NMR signal (record resonance energy). A typical NMR spectrometer is outlined in Figure 11.3.

The earliest form of NMR spectroscopy relied on CW methods, where experiments were conducted with the RF field present throughout, bringing nuclei with different chemical shifts into resonance by either keeping the electromagnetic frequency fixed and sweeping the magnetic field, or vice versa. However, the origin of pulse methods including FT-NMR soon superseded CW methods. Pulse methods depend on applying a short intense burst of RF radiation to a sample to set nuclear moments spinning around the line of the magnetic field in a motion called *precession*. As such, each nucleus precesses around the static field direction at a frequency called the *Larmor frequency*. At the end of the pulse, the oscillating magnetization is then detected as a function of time and processed by a computer to give rise to an NMR spectrum.

To observe magnetic resonance, a sinusoidally oscillating electric current is passed through a coil wrapped around a sample sitting inside a magnet. Electrons circulating in the coil generate an oscillating electromagnetic field and, providing the frequency is satisfactory, the magnetic component excites the sample. The

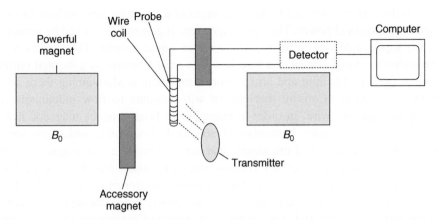

Figure 11.3 Diagrammatic representation of an NMR spectrometer. (Adapted from Sheehan, Physical Biochemistry, 2000.)

coherent RF field then causes the magnetic moments of the nuclei to precess in phase around the intense static field at their individual resonance frequencies. The oscillating magnetization then induces an alternating current in a nearby receiver coil and this can be amplified and presented as the NMR signal.

Transmitter

As the name suggests, the transmitter is the part of the spectrometer that generates and transmits RF pulses to the sample. An RF waveform generator (source) emits a continuous voltage, oscillating sinusoidally at a desired frequency. Output from the RF generator is converted into so-called *phase-shifted RF pulses* by a switch (or gate) opened and closed by a computer-controlled pulse generator. The low level pulses then input to an RF amplifier that serves to boost the pulses sent to the sample in the NMR probe, housed in the magnet. As such, the computer-controlled transmitter ultimately generates pulses of required duration and power needed for the specific NMR analysis.

Magnet

NMR magnets are one of the most expensive parts of NMR spectrometer systems. Modern NMR spectrometers use *superconducting magnets* which are electromagnets comprising a coil of superconducting wire through which an electric current passes. When cooled to temperatures approaching absolute zero ($-273.15\,^{\circ}$C), superconducting wire has a resistance approximately equal to zero; so once current flows in the coil, it continues to do so, generating a magnetic field without further consuming electrical power. The low temperature is generally achieved by immersing the superconducting wire in a container of liquid helium. This container is surrounded by liquid nitrogen, a so-called *heat shield*, acting as a thermal buffer between the liquid helium and air (at room temperature). Maintaining these conditions is essential to ensure that current will continue to flow indefinitely in the superconducting wire. In order to maintain the homogeneous magnetic field, the spectrometer contains a number of so-called *shim coils*, which are small coils of wire placed around the sample that serve to produce tiny magnetic fields which cancel out *field gradients* (*inhomogeneities*) which may be produced by the *solenoid* (main magnetic field). Most modern spectrometers control the shim coils automatically by computer-generated *algorithms*. While the superconducting magnet produces a very stable magnetic field, there will nevertheless be some *drift* in the field which is corrected by so-called *field frequency lock*. The B_0 field is monitored via the NMR frequency of a reference compound, usually the ^{2}H signal of a *deuterated solvent*. The field frequency lock is tuned to the deuterium NMR resonance frequency, where the instrument constantly monitors the resonance frequency of the deuterium signal, and makes minor changes in the magnetic field to compensate for the drift, thus keeping resonance frequency constant. Conveniently, the deuterium signal can come from a deuterium solvent used to prepare the sample.

Probe

The probe is a cylindrical metal tube which (i) holds and spins the sample, (ii) sends RF energy into the sample and couples the RF field to the spins, and (iii) detects the NMR signal arising from the sample. This cylindrical tube containing the sample *spinner*, electronics controlling temperature, and RF coil (and sometimes gradient coils), is inserted into the *bore* (centre) of the magnet. The sample spinner rotates the NMR sample tube about its axis in order to correct inhomogeneities and generate a narrow NMR line-width on the spectrum. Sample temperature is monitored with the aid of a *thermocouple* and electronic adjustments maintain the temperature by either increasing or decreasing the temperature of the gas passing over the sample. The *RF coil* of wire (or foil) is the crucial component of the probe. This coil lies around the sample and receives transmitter pulses, and the arising alternating current in the coil generates a magnetic field of the same frequency and phase as the transmitter. This RF field excites nuclear magnetization, inducing an oscillating voltage in the coil, the NMR signal, which is sent to the receiver. The RF coil resonates in the system at the specific Larmor frequency of the nucleus being examined with the NMR spectrometer. In other words, adjustments to the RF coil tune the instrument for the specific sample under investigation. *Gradient coils* are different from the RF coil in that they produce gradients in the magnetic field required for gradient-enhanced spectroscopy, diffusion measurements and NMR microscopy. As noted above, not all probes have gradient coils, and thus not every NMR spectrometer has the ability to drive these coils.

Receiver

The NMR signal from the probe is sent to the receiver, where it is amplified, and the RF signal then inputs into a detector which mixes this signal with a reference voltage. Usually the reference voltage is of the same frequency as the pulses used to excite the spins (i.e. those emitted from the RF generator), and mixing subtracts the reference frequency from the NMR signal to produce an *audiofrequency* (AF) voltage. The AF voltage then passes from the detector to an AF amplifier and from there to an analogue-to-digital converter linked to a computer for final signal processing.

11.3   Established and emerging applications of NMR

Nuclei such as ^{1}H or ^{13}C resonate at a specific frequency in a magnetic field of given strength (measured in *tesla* (T)). NMR instruments are often named on the basis of the particular magnetic field strength they generate, for example, a 21 T magnet is commonly referred to as a 900 MHz magnet (see Figure 11.4). Other commonly encountered NMR instruments operate at 400, 500 or 600 MHz, but historically other less powerful magnets have been used (e.g. 250 MHz). These NMR instruments have allowed in-depth analysis of biomolecules, yielding more information than could be generated from other forms of spectroscopy.

NMR spectroscopy, and in particular ^{1}H or ^{13}C NMR, has provided detailed information on structure and function attributes of major classes of biomolecule. Indeed, the study of peptides and proteins has lead to major developments and new applications of NMR. Notably, NMR has enabled researchers to follow or trace biochemical pathways and has emerged as an important technology in the analysis and development of new drugs.

Historically NMR spectroscopy has been utilized primarily for *in vitro* bioanalyses, but NMR can also be applied to the study of chemical reactions in living organisms ranging from cells to whole body measures. ^{31}P NMR is especially useful for *in vivo* applications such as measurement of energy metabolism in heart muscle, benefiting from the fact that ^{31}P resonance is relatively easy to observe and generates spectra within seconds. However, while ^{1}H, ^{13}C and ^{31}P NMR have

Figure 11.4   **Photograph of a modern NMR spectrometer.**

applications of medical importance, the introduction of MRI has greatly extended
the power of magnetic resonance in diagnostic medicine.

11.4   Principles and uses of MRI

Magnetic resonance imaging (MRI) builds on the principles of NMR. Indeed, the
principles of NMR discussed in earlier sections directly apply to MRI. However,
from the outset it is important to note that the main distinction is that MRI relies
solely on the magnetic properties of the hydrogen atom to produce images and thus
is comparable to ^{1}H NMR, but quite different in its scope as it does not encompass
molecules other than hydrogen. Of course, this is not a limitation, as all biological samples and tissues are constructed from hydrogen-containing molecules.
MRI was formerly known as *magnetic resonance tomography* (MRT) and the
original terminology used in medicine was 'NMRI'. However, the word nuclear
was quickly dropped as it has negative connotations and importantly radiation
exposure is *not* one of the safety concerns of this method. As this was primarily

done to avoid patients from falsely associating MRI with radiation exposure, many scientists still use 'NMRI' when they discuss devices that work on the principles of NMR.

As with NMR, the physics underlying MRI is extremely complex and thus falls outside the remit of this textbook, but the following provides a brief summary of the main principles and uses of this increasingly widespread and powerful imaging technology.

A single proton is the heart of a hydrogen atom and, as discussed earlier, it is a spinning charged particle with inherent magnetic properties. Like NMR, MRI relies on placing H atoms in a strong external magnetic field, but unlike NMR, medical MRI does not rely on patient samples, rather the entire person is the sample. As such, MRI is a medical doctor's diagnostic dream, in that it essentially permits noninvasive surgical examination; that is, looking inside the body without the need for the patient to go 'under the knife'. However, MRI is not cheap and thus is restricted in availability, and it does not allow surgeons to answer all the questions they may have; so it should be considered a useful noninvasive tool rather than replacement for more traditional surgical examination (see Figure 11.5).

MRI is particularly useful for anatomical examination, and is often seen on television as the futuristic means by which doctors scan patients to see internal structures such as the brain or other localized areas. Historically, MRI studies mainly focused on the head and extremities (e.g. hand or leg), as imaging of moving organs, particularly the heart and lungs, was difficult to resolve and thus more challenging experimentally. However, MRI technology has continued to evolve rapidly, and modern MRI equipment does not suffer from the limitations of earlier instruments.

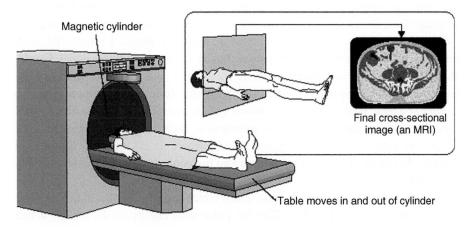

Figure 11.5   Diagram of an MRI scanner.

Medical MRI relies mainly on the relaxation properties of excited hydrogen nuclei. In MRI the patient is placed in a powerful superconducting magnet. This magnet provides a uniform field, commonly 0.3–3 T, but this may reach 20 T or higher in research instruments (for a useful indication as to how powerful this actually is, the Earth's magnetic field is around 0.00005 T). In this uniform magnetic field, the hydrogen nuclei within the patient align either in a parallel or anti-parallel direction (see earlier) to this external magnetic field. For simplification, it is taken that the majority of the ^{1}H nuclei align in a parallel direction and therefore will be in the same direction as the applied external field.

In MRI, the tissue is exposed to a magnetic field (typically 1.5 T), and the ^{1}H nuclei (or magnetic dipole moments) spin around the line of this field in a motion known as *precession*. The frequency of precession in an H atom in a given magnetic field is predefined and referred to as the Larmor frequency, which changes in proportion to the magnetic field strength and is typically in the order of 10 MHz (radiofrequency). The tissue in the magnetic field is then briefly exposed to RF pulses (at the Larmor frequency), which are applied perpendicular (i.e. at right angles) to the original magnetic field. This results in the magnetically aligned hydrogen nuclei assuming a temporary *non-aligned high-energy state;* that is, these hydrogen nuclei turn in a direction at right angles to the original magnetic field.

Images are constructed based on the movement of the ^{1}H nuclei, and for this different magnetic gradients can be applied to the principal axis of the patient so the subject is imaged in x, y and z directions, although MRI allows more flexibility. These different magnetic gradients allow construction of 1D, 2D and 3D image data from the patient. These images are created from MRI data using *discrete Fourier transform* (DFT). Fourier transform is a linear, reversible transform that maps an image from the space domain to the frequency domain. Converting an image from the spatial to the frequency (Fourier) domain helps in assessing the spectral content in energy distribution over frequency bands. Commercial medical scanners can piece together images based on the two-dimensional Fourier transform (2DFT) or *spin-warp* method. This is a complex process, but basically the *imaging pulse sequences* are composed of the following common steps: (i) *excitation (slice selection)*, (ii) *evolution (phase encoding)*, (iii) *detection (frequency encoding)*, and (iv) *image reconstruction*. Information is encoded by three parameters (amplitude, frequency and phase), where in 2D imaging, localization is controlled by the frequency and phase and the contrast is given by the intensity (amplitude).

MRI *contrast* is dependent on *relaxation*, which occurs on cessation of the brief exposure time to the RF pulses. The RF pulse adds energy to the nuclei, which is absorbed, and during relaxation the RF pulse energy transferred to the nuclei is dissipated to the surrounding environment (or *lattice*) and the nuclei relax and realign. The rate of emission of this energy can be recorded and gives important information about the environment. The process by which nuclei transfer energy to the lattice is denoted as T_1, which is also commonly referred to as *spin–lattice relaxation time* (T_1 recovery). This realignment or recovery of nuclear spins with the original magnetic field is called *longitudinal relaxation*. The other change that happens on removal of the RF pulse is referred to as T_2, the so-called *spin–spin relaxation time* (or *transverse relaxation*), which occurs when precessing ^{1}H nuclei spins *dephase*. This is essentially a form of decay which happens as the small magnetic fields of neighbouring nuclei interact, resulting in a loss of transverse magnetization and hence signal. A related method is so-called T2* imaging, where a *spin-echo technique* (so-called spin-echo pulse sequence) is used to compensate for local magnetic field inhomogeneities. As such, T2* imaging combines two effects: (i) T_2 decay, and (ii) dephasing due to magnetic field inhomogeneities, and in essence MRI uses this spin-echo sequence in generating images.

A spin-echo sequence uses a 90° RF excitation pulse followed by one or more 180° rephasing RF pulses, and each of the rephasing pulses generates a separate spin echo that can be received by the coil and used to generate an image. The time between each 90° RF excitation pulse is referred to as the *repetition time* (or TR) and the time between the start of the 90° RF excitation pulse and the peak of the spin echo is called the *echo time* (or TE). The MRI signal and thus image contrast relies on a number of factors which include: (i) proton density (number of protons per unit volume of tissue; no relaxation time); (ii) T_1; (iii) T_2; (iv) T2*; (v) chemical environment of H atoms (e.g. in water or fat); (vi) flow, that is blood vessels or cerebospinal fluid (CSF). An MRI image has contrast if there are areas of high (white), intermediate (shades of grey) and low (black) signal. High signals result from large components of *transverse magnetization* resulting in *brightness*. Images obtain contrast mainly through T_1, T_2 and proton density. There are two main extremes of MRI contrast: fat which has short T_1 and T_2 times, and water which has long T_1 and T_2 times, as energy exchange in water is less efficient than fat. Given this, images are obtained following so-called *weighting*. T_1-weighted images have spin-echo pulse sequences with short TR (exaggerating T_1) and short TE (diminishing T_2), with resulting images characterized by bright fat and dark water. T_2-weighted images have spin-echo pulse sequences with long TE (exaggerating T_2) and long TR (diminishing T_1), and resulting images are characterized by bright water and dark fat. Proton-density weighted images have spin-echo pulse sequences with long TR and short TE (i.e. T_1 and T_2 diminished), where areas of high proton density are bright, and areas of low proton density are dark. To add variability and complexity, an alternative form of echo pulse sequence exists called *gradient-echo pulse sequence*, and in contrast to spin-echo pulse sequence which has a fixed 90° RF excitation pulse, the gradient-echo pulse sequence uses a variable (<90°) RF excitation pulse.

In practice, T_1-weighted images have excellent anatomical definition but are less effective in imaging pathologies, whereas T_2-weighted images are highly effective in imaging/detecting pathologies. To give an example of how such images appear, T_1-weighted brain images allow nerve connections (white matter) to appear white, and dense tissue comprising neurons (grey matter) to appear grey, while CSF appears dark. The contrast in T_1 imaging may be changed or reversed using T_2 or T2* imaging, but proton-density weighted imaging provides little contrast in normal subjects. Notably, functional information is encoded within T_1, T_2 or T2*, which is often termed functional magnetic resonance imaging (fMRI) and can describe parameters such as vascular perfusion, detailed in the next section.

Using MRI and safety issues

MRI is a powerful imaging technique which has no directly reported hazardous biological effects. However as this technique relies on a very strong magnetic field and use of RF pulses, there are a number of safety issues when using an MRI scanner. Most of these relate to various implants that may be electrically-active, which include *cardiac pacemakers*. Such implants are generally considered to be *contraindications* to the use of MRI scanning and there have been some cases of heart arrhythmias in patients with pacemakers who have been exposed to MRI scanners, largely believed to result from effects on the electrodes or wiring. Ferromagnetic materials or metallic implants may also potentially be moved or disrupted when a patient is exposed to the strong MRI magnetic field, and of these the most common issues related to metal fragments in the eye (e.g. workplace injury) and surgical prostheses (e.g. aneurysm clips). This has resulted in more widespread use of titanium-based materials. Given this, implants are often classified as (i) *MRI safe* where the implant is unlikely to cause injury by moving, heating or exerting electrical effects during MRI scanning; and (ii) *MRI compatible*, where implants are both MRI safe and will not degrade the image.

Other key safety points are given briefly below:

Projectiles: many metal objects can become airborne missiles in the presence of a strong magnetic field, being attracted to the centre of the magnet. For example, a simple paperclip or hairpin can reach a velocity of 40 miles per hour in the presence of a 1.5 T magnet. Given this, it is easy to see why there are reported cases of MRI-related injury and death.

Claustrophobia: typically MRI scanners require the patient to lie in a long narrow tube, placing them in the centre of the magnet. As MRI scanning can take some time, people with even mild claustrophobia can find it a difficult experience and sedation/general anaesthesia may be required.

Noise levels: MRI scanners used rapidly switching magnetic gradients and as the coils expand and contract, high auditory noises and vibrations, which can reach 130 dB (the sound intensity of a jet engine), can be emitted, so appropriate ear protection is required.

Peripheral nerve stimulation: the rapidly switching magnetic gradients can also stimulate muscles and peripheral nerves. This may thus be reported as a twitching sensation, particularly at the body extremities. Guidelines have been introduced to limit the rate of switching, aiming to avoid this effect.

RF burns: although very rare, exposure to the MRI scanner's powerful radio transmitter has been reported to cause burns, and the risk of hyperthermia is increased in children or the elderly.

Pregnancy: while there is no evidence of harmful effects of MRI on a fetus, and fetal scans can produce useful clinical diagnostic information, precautionary guidelines have been introduced recommending that MRI scanning is limited to cases of absolute necessity in pregnant patients. In addition, the *contrast agents* (e.g. gadolinium compounds) can enter the fetal bloodstream, and thus guidelines suggest that their use is avoided.

11.5 MRI as a principal diagnostic and research tool

MRI is a versatile method and in addition to uses in diagnostic medicine, it is important to note that while MRI is best known as a medical tool, this technique has other useful scientific applications. As an example, geologists have used MRI to examine the composition of geological structures, and MRI has also been exploited in assessment of produce and timber quality. However, as noted earlier, MRI scanners cost millions to buy and thousands to run and maintain, so remain a luxury big-budget item.

Clinically, MRI is used as a diagnostic tool, which is useful for distinguishing pathological from normal tissue (e.g. a solid brain tumour–see Figure 11.6). Typically, MRI examinations consist of multiple sequences selected based on the information required, and images arising are interpreted by the clinician. One major advantage of MRI is that, unlike CT scans and X-rays, it does not use ionizing radiation. MRI has evolved in such a way as to allow specialized scans of various body tissues, which can include the use of contrast materials to help enhance image resolution.

While T_1-weighted and T_2-weighted images provide useful clinical information, sometimes it is important to employ more specialized image acquisition

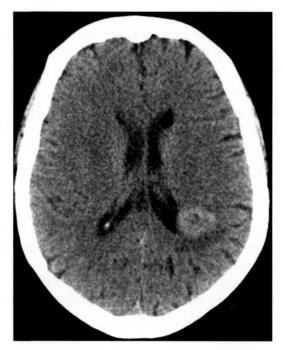

Figure 11.6   MRI scan revealing a brain tumour.

methods (e.g. fat suppression and chemical-shift imaging), and administration of contrast agents can also help resolve particular regions. Contrast agents (Figure 11.7) include water or diamagnetic, paramagnetic and superparamagnetic agents. Water has been used for imaging of the stomach and small intestine, and the *diamagnetic agent* barium sulfate may be useful for gastrointestinal (GI) tract imaging. The *paramagnetic agent* gadolinium enhances T_1-weighted images, and as it enters tissues and fluids it brightens well-vascularized tissue, thus making it useful for detection of tumours, lesions, and vascular perfusion (e.g. in stroke), particularly in the brain, spine and musculoskeletal system. *Superparamagnetic contrast agents*, including iron oxide nanoparticles, have proven particularly useful for imaging liver and the GI tract. Normal liver, which retains the agent, appears dark on T2*-weighted images and thus can be differentiated from abnormal regions, such as scars and tumours.

Major diagnostic applications of MRI: specialized scans

Magnetic resonance angiography (MRA): This technique is used to image the arteries and blood flow, and using computer reconstruction it is even possible to display blood vessels in 3D. Thus, MRA is useful for evaluation of abnormal

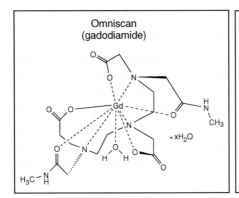

Figure 11.7   **Examples of major contrast agents used in MRI.**

narrowing (stenosis) or dilation (including aneurysms) of the blood vessels, and is often used to evaluate major arteries. Images can be enhanced by the use of contrast agents (e.g. gadolinium compounds) or by so-called *flow-related enhancement*, where blood flow causes an increased signal. It is also important to mention *magnetic resonance venography* (MRV), which uses a similar approach to image veins.

Diffusion MRI: This technique relies on the diffusion of water molecules throughout the body and its tissues. Biologically, diffusion refers to a *random motion* (called *Brownian motion*) of molecules driven by thermal energy. In order to make MRI sensitive to diffusion it is necessary to add gradient pulses to spin-echo sequences by so-called diffusion-weighted imaging (DWI). With DWI, signal loss occurs in areas of high diffusion rates and is thus useful in detecting the pathological changes occurring in the lesions resulting from ischaemic stroke. The DWI enhancement occurs very quickly (within minutes) compared with CT, which can take many hours to detect the same lesion.

Perfusion MRI: In contrast to diffusion MRI that examines the movement of fluid through tissues, perfusion MRI determines blood flow on the basis of measures of blood volume. *Perfusion-weighted imaging* (PWI) can determine cerebral tissue perfusion, including measures of blood flow and transit time, to create maps of relative cerebral blood flow and detect areas of reduced perfusion.

Functional MRI (fMRI): Perhaps confusingly this technique is restricted to measuring signal changes in the brain, reflecting alterations in neural activity. An increase in neural activity results in an increased demand for oxygen, which is provided by the bloodstream, and this is detected by *blood oxygen-level-dependent* (BOLD) imaging. As such, BOLD is a non-invasive method to investigate cerebral physiology and pathology. The MRI signal can also be weighted by *cerebral blood flow* (CBF) or *cerebral blood volume* (CBV), which may further expand the applications of fMRI.

Magnetic resonance spectroscopy (MRS): This technique is also referred to as magnetic resonance spectroscopy imaging (MRSI) or *volume-selective NMR spectroscopy*, which basically combines MRI with NMR spectroscopy. Thus MRS allows conventional MRI study of a body region, while NMR spectroscopy provides information on the biomolecules in that same region. This is a powerful tool, which can be used to provide spectral data on metabolites including: lactate, choline, creatinine and myoinositol.

Key Points

- Nuclear magnetic resonance lies at the heart of both NMR spectroscopy (biomolecule identification/characterization) and MRI (non-invasive imaging).

- ESR is an alternative to NMR spectroscopy, while the major alternative to MRI imaging is CT, otherwise known as *computer axial tomography* (CAT/CT scan).

- NMR depends on magnetic properties of atoms and their nuclei, where magnetic nuclei orient/align parallel to (with) or anti-parallel to (against) the external magnetic field.

- As nuclei absorb electromagnetic energy they are promoted from a lower to higher energy state (resonance) and return to the lower energy state (relaxation), generating a NMR spectrum.

- Three essential requirements for NMR recordings are: an intense static magnetic field with defined field strength, a high frequency transmitter, and detector.

- NMR instruments are often named on the basis of the particular magnetic field strength they generate, for example, a 21 T magnet is commonly referred to as a 900 MHz magnet.

- MRI relies solely on the magnetic properties of the hydrogen atom to produce images (comparable to ^{1}H NMR), representing 'non-invasive surgical examination'.

- MRI is used in clinical diagnosis, distinguishing pathological from normal tissue (e.g. a solid brain tumour), and unlike CT and X-rays does not rely on ionizing radiation.

- MRI can include the use of specialized image acquisition methods, and administration of contrast agents can also help enhance images and resolve particular regions.

- Major diagnostic applications of MRI in specialized scans include: MRA; diffusion MRI; perfusion MRI; fMRI; and MRS.

12 Bioanalytical approaches from diagnostic, research and pharmaceutical perspectives

Earlier chapters have outlined the fundamental principles and applications of core bioanalytical techniques commonly utilized in the life and health sciences. The principal aim of this chapter is to give insights into the use of these tools in various different settings to illustrate the power and diversity of core technologies in diagnostics, research and development. At the heart of bioanalysis is the application of science to improve the ability in identifying, characterizing, quantifying, and ultimately enhancing knowledge and understanding of molecules in artificial and natural environments. Historically, bioanalytical chemists have often focused on the use of a single type of instrument but, increasingly, academic, clinical and industrial fields are harnessing the potential offered by the combined use of multiple technologies to develop new methods with enhanced specificity and sensitivity. During the 1900s and into the 2000s, instrumental analysis has become more dominant. Indeed, in the last 30 years the combined approach to analytical chemistry has progressively shifted in focus from inorganic or small organic molecules to larger and often more complex biomolecules. This has allowed a transition from largely academic applications to answer key biomedical, industrial and environmental questions. The following sections provide a brief overview of some focus areas of bioanalytical chemistry in life and health sciences.

Understanding Bioanalytical Chemistry: Principles and applications   Victor A. Gault and Neville H. McClenaghan
© 2009 John Wiley & Sons, Ltd

Learning Objectives

- To introduce the emerging fields of clinical genomics, proteomics and metabolomics.

- To highlight the use of bioanalytical approaches to clinical diagnosis and screening.

- To provide a brief overview of how bioanalytical chemistry can drive fundamental research and development activities, and commercial drug discovery and clinical development.

- To give examples of how translational research can result in pharmaceutical products from natural sources, and the emergence of biologics as therapeutic agents.

- To appreciate ongoing developments and future perspectives in genomics, proteomics, personalized medicine, and identification of new therapeutic applications for known drugs.

12.1 Clinical genomics, proteomics and metabolomics

There has been an explosion in the use of the term 'omics', describing the latest and most innovative approaches to understanding the biomolecular basis of health and disease. The term 'omics' comes from the word *genome*, and as genome refers to the complete genetic constituents of an organism, in general the use of the suffix '–omics' is largely taken to refer to a totality of some kind and is popularly used by *molecular biologists* and *bioinformaticians* or *system biologists*. Given the very large amounts of data being generated by scientists working in the 'omics', *systems biology*, which provides the 'number-crunching' tools to make sense of this data, is becoming an increasingly important part of these fields of study. Currently, the most popular fields are *genomics* (study of the 'genome' – genes), *proteomics* (study of the 'proteome'–proteins) and *metabolomics* (study of the 'metabolome' – metabolites) (see Table 12.1). These areas are vast and, as such, detailed discussion is beyond the scope of this book, however, some relevant clinical illustrations are given below.

Table 12.1   **Examples of new and emerging 'omics'**

	Brief description
Exposomics	Describes the total environmental exposure(s) of an organism (exposome).
Glycomics	Total complement of glycans (type of carbohydrate molecule) in an organism (glycome).
Interactomics (systems biology)	All of the molecular interactions in an organism (interactome).
Lipidomics	Total complement of lipids in an organism (lipidome).
Neuromics	Study of the complete neural biology of an organism (neurome).
Omiomics	Totality of all 'omes' which will be the future catalogue of all omics studies (omeome).
Peptidomics	Total peptides in an organism (peptidome).
Phenomics	Describes the organism – the phenotype of an organism (phenome; including any mutations)
Physiomics	Describes the study of the complete physiology of an organism (physiome).
Transcriptomics	Total messenger RNA (mRNA) of an organism (transcriptome).

Please note: Many of these terms may also be used to describe isolated tissue and/or cell.

Clinical genomics

Genomics can generally be considered as a comprehensive analysis of gene expression by assessing *relative* or *semi-quantitative* amounts of RNA in biological samples. This field extends to cover *mutations*, *deletions* and other changes in genes that may affect their expression. Since the publication of the complete genome of the nematode worm, *Caenorhabditis elegans* (or *C. elegans*) in 1998, major efforts have focused on other species, including the Human Genome Project – with genomic sequencing largely completed in 2006. Availability of human genomic sequence data coupled with advances in bioanalytical tools such as reverse transcription-polymerase chain reaction(RT-PCR), low-density arrays and high-density DNA microarrays (advanced electrophoretic separation technique), facilitates the study of complex diseases such as cancer through generation of large amounts of data for clinical research. This has helped drive knowledge and understanding of the molecular mechanisms of cancer biology and notably: (i) discovery of new biomarkers for early cancer detection; (ii) tumour classification; (iii) metastasis prediction; (iv) response to therapy; and (v) disease prognosis. New and developing genomic tools including *DNA microarrays* allow the data capture and analysis of many thousands of individual genes and their interaction, to generate a global picture of, for example, cancerous tissue. The ultimate challenge in cancer is to detect the condition as early as possible in order to maximize therapeutic and clinical outcomes, and particularly exciting in this regard is the study of cancer biomarkers. For optimal therapy, highly sensitive and specific biomarkers are required for early diagnosis, and while none of the FDA-approved biomarkers (such as prostate specific antigen, PSA; and alpha-fetoprotein, AFP) are presently used in standard clinical practice, with time, identification of more specific biomarkers may lead to universal adoption in clinical screening programmes.

Clinical proteomics

While genomics is a powerful tool, it currently remains hypothesis-driven, in that changes in gene expression may not always translate into detrimental changes in expressed proteins. As such, proteomics has been proposed as the logical next step in understanding the genetic basis of biological processes and disease. Given this, the complementary study of genomics and proteomics should not be underestimated, as it can balance the strengths and weaknesses of each individual approach. Proteomics can generally be considered as a comprehensive analysis of proteins in biological samples and, somewhat confusingly, peptide analysis may also be considered under this umbrella term, despite the increasing popularity of the field of 'peptidomics'. The field of proteomics is complex, including detection, identification and quantification of proteins, together with their interactions, regulation and modifications. To highlight the complexity, while there are around 30 000 human genes in the genome, there are probably 35 times more proteins in the human proteome. The proteome reflects both intrinsic genetic programming of an organism and environmental influences and, like clinical genomics, a primary goal of clinical proteomics is the identification of early disease biomarkers and therapeutic targets. Exciting opportunities also exist to match disease biomarkers with diagnostic imaging tools to improve disease detection and monitoring of recurrence of, for example, gastrointestinal diseases. Important bioanalytical tools for proteomic analysis include advanced chromatographic and electrophoretic separation techniques, mass spectrometry (MS), immunoassays and tissue/protein microarray platforms. Tissue/protein microarrays and biochip technologies offer great potential for high-throughput analysis of very many proteins that may be altered in a particular disease state.

Clinical metabolomics

The metabolome is essentially a snapshot representation of all metabolites in a cellular/tissue system at a distinct point in time. Whereas the analysis of genes and proteins relies on knowledge of genome sequence or protein blueprint, analysis of metabolites in an unknown sample is far from straightforward. Complexities lie in the lack of simple automated bioanalytical techniques to measure the many hundreds or thousands of labile and chemically diverse metabolites that may be present in a sample. Given that the metabolome will vary with pathology, developmental, environmental, and other factors that may influence metabolism, methods are required to 'stop metabolism' before measurement, which currently precludes effective reproducible and robust quantitative analysis. Analysis of a metabolome can focus on certain themes such as: (i) analysis of metabolite targets; (ii) profiling of metabolites; and (iii) fingerprinting and footprinting of metabolites. These themes utilize core technologies including advanced chromatographic separation MS tools (for example gas chromatography mass spectrometry (GC-MS), liquid chromatography mass spectrometry (LC-MS), capillary electrophoresis mass spectrometry (CE-MS) and ultra-performance liquid chromatography mass spectrometry (UPLC-MS)) or spectroscopic techniques (for example Fourier transform infrared (FT-IR) and nuclear magnetic resonance (NMR)), and solid databases are essential for appropriate data analysis and interpretation. Despite inherent limitations, many investigators have seen the value of metabolomics in understanding metabolism in different species and how this knowledge can project into a clinical setting. For example, metabolomics offers great potential in probing the molecular pathophysiology (including metabolic syndrome, Type 2 diabetes, cardiovascular disease and obesity), and providing discovery platforms for biomarkers associated with health and disease.

12.2   Clinical diagnosis and screening

Most current and future tests used clinically for diagnosis and screening are based on known biomarkers for disease. These biomarkers may reflect defects or mutations in one or more genes and/or proteins, which may be inherited. While major diseases such as diabetes appear polygenic in origin, there are still over 10 000 human diseases that are understood to arise as a result of single gene defects. The cheapest genetic screen is the 'family history' which may provide a fundamental indication of risk of birth defect, and while there is around a 1% risk of a newborn in the general population inheriting a *single gene disorder*, knowledge of family history could reduce this risk by as much as 25 or 50% for *autosomal recessive* disorders or *autosomal dominant* disorders, respectively. Advances in molecular genetics provide the diagnostic and predictive tools for a more robust analysis of genetic inheritance, which may be categorized as: (i) *direct* – detection of specific mutation responsible for disease or disease risk; (ii) *linkage-based* – rely on use of genetic markers allowing tracking of inheritance; and (iii) *genomic testing* – analysis of multiple variants or products of genes (providing *genomic fingerprint*). There are four primary indications/applications of genetic testing, namely: (i) *diagnostic testing* – aims to provide precise diagnosis of individual with signs/symptoms of disease; (ii) *pre-symptomatic testing* – aims to assess individuals with family history who do not yet show signs/symptoms of disease; (iii) *pre-dispositional (susceptibility) testing* – applied to multi-factorial disorders arising from collision between genes and/or environment; and (iv) *pharmacogenetic testing* – involves examination of genes that control drug metabolism/activity to attempt to enhance 'personalized' drug efficacy and reduce side effects in individuals. When assessing the use of a genetic test it is important to consider its *analytical validity* (likelihood reported test results are correct), *clinical validity* (degree to which test correctly assesses risk) and *clinical utility* (degree to which test links with clinical management to improve outcome). In this regard, it is important to consider the ethical, legal and social concerns inherent in genetic testing, which are hotly debated.

There are a number of approaches to clinical diagnosis and screening based on genomic, proteomic and metabolomic analyses of samples taken from individual persons, tissues or cells. These techniques cover the spectrum of technologies described in this book, directing current and future treatment and even prevention of disease. Current routine diagnostics in *clinical chemistry* have greatly benefited from the emergence of advanced automated tools encompassing bioanalytical instruments, specimen-processing equipment and laboratory assays. Such tools include *automated high-throughput clinical analysers* which can run a battery of routine and advanced tests; for example on blood samples to generate data on target parameters including glucose and iron status. These clinical analysers, including the popular Roche/Hitachi systems, allow analysis of samples (including blood serum/plasma, urine, cerebrospinal fluid), encompassing both classical and special clinical chemistries, and homogeneous immunoassays. Clearly, this area of bioanalysis is huge and steadily growing as research yields new tools and approaches to clinical diagnosis and screening, where biomarkers validated in large clinical studies are often of central importance. The following example illustrates how this field is emerging to utilize other genetic approaches for future diagnosis of disease prediction and/or progression.

Cystic fibrosis: Of the single gene defect disorders, particular attention has been given to cystic fibrosis (CF), a so-called autosomal recessive disorder (incidence

~1:500 births) caused by mutation of the cystic fibrosis transmembrane conductance regulator (*CFTR*) gene located on human chromosome 7. Inheritance of CF follows a Mendelian pattern (see Figure 12.1), where if a male carrier and female carrier (each having one copy of the mutated CTFR gene) have a child, there is a one in four chance it will be affected by CF. Notably if these parent carriers have another child it also has the same one in four chance of being affected (Figure 12.1). The CTFR gene encodes a protein that acts as a chloride ion (Cl^-) channel and modulator of sodium (Na^+) transport into cells of many organs, notably the lungs. In persons affected by CF, thick and sticky mucus is produced in the lungs causing congestion and an increased chance of respiratory infections. Over 1000 mutations of CFTR have been detected, but in ~70% of cases the mutation involves deletion of the amino acid phenylalanine at position 508. Historically, the gold-standard diagnostic test for CF relied on measurement of sweat electrolyte (normally Na^+ and Cl^-) concentrations. While this test picks up most affected individuals, modern genetic approaches can be utilized to cover all subpopulations and additionally facilitate screening, particularly of newborns. Considerable debate has focused on inclusion criteria for CF mutations, which currently recommend screening for 23 of 25 known mutations of CFTR, which accounts for >90% of detectable mutations. However, commercially

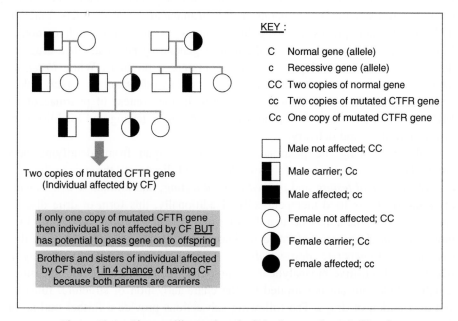

KEY :

C   Normal gene (allele)
c   Recessive gene (allele)
CC Two copies of normal gene
cc Two copies of mutated CTFR gene
Cc One copy of mutated CTFR gene

☐ Male not affected; CC
◼ Male carrier; Cc
■ Male affected; cc
◯ Female not affected; CC
◑ Female carrier; Cc
● Female affected; cc

Two copies of mutated CFTR gene
(Individual affected by CF)

If only one copy of mutated CFTR gene then individual is not affected by CF BUT has potential to pass gene on to offspring

Brothers and sisters of individual affected by CF have 1 in 4 chance of having CF because both parents are carriers

Figure 12.1   **Diagram illustrating the inheritance of cystic fibrosis.**

available panels for screening may be significantly expanded beyond these mutations, based on the hypothesis that testing for more mutations can substantially
enhance detection in Hispanic and African-American population groups. CF may
be viewed as a model of success, and as more genes become amenable to *prenatal
testing* it is likely that other diseases with difficult prognoses or high mutation
frequencies may be considered for screening. Candidates include spinal muscu-
lar atrophy (SMA), which is a common autosomal recessive disorder (incidence
1:10 000 live births) characterized by severe progressive neurodegeneration, with
no available treatment. This further highlights the ethical and social considerations
and implications of genetic testing.

12.3   Research and development

Earlier sections have focused on current and emerging approaches to bioanalytical
methods and tools from a diagnostic perspective. However, bioanalytical chemistry
drives fundamental research and development (R&D) activities in academic and
commercial settings. Academic research encompasses the entire range of available
bioanalytical instrumentation, and often technologies evolve and expand on the
basis of inventive ways of addressing fundamental or progressive scientific questions. This process feeds method development and validation that also informs
industry, which, in turn, may translate outcomes of primary exploratory studies
through additional R&D to generate new instrumentation and products. Inherent
in the development process is taking this primary experimental research to another
level of scale or complexity, with the aim of generating new candidate drugs or
diagnostic agents, technologies, or other commercial products. While *Research*
largely refers to experimentation and discovery, *Development* is largely taken
to be the process of exploitation of research. In the context of pharmaceutical
R&D, development starts with proof-of-concept, safety testing and establishment
of effective dose and delivery.

Challenges facing the pharmaceutical industry span from identifying new
patentable agents and targets for drug design and discovery, through to the process of *preclinical testing* and *phased clinical testing*, and ultimately assessments
of *validated drugs* used therapeutically. Traditionally, this form of drug discovery and development has been a *linear process*, but this may well shift to a
more *integrated process*, informed by knowledge of the molecular basis of disease in order to generate '*designer drugs*'. Drug discovery utilizes a vast array of
selective and sensitive bioanalytical technologies, facilitating important measures
which include, but are not limited to, the characterization of structure, function
and distribution of drugs. The following subsections provide a further outline of
drug discovery and development with reference to bioanalytical technologies.

Drug discovery

Early drug discovery was often by serendipity, where chance observation in 'human guinea pigs' often informed clinical practice. Sometimes this involved discovery of an active ingredient from natural plant products (e.g. aspirin, or salicylic acid, was derived from an active extract from white willow bark), but the term *drug discovery* also includes the process by which drugs are designed. The process of drug discovery generally involves: (i) identification of candidate agent; (ii) synthesis of agent (or derivative); (iii) structural characterization; (iv) biological screening; and (v) determination of therapeutic action. These initial exploratory investigations using chromatographic (e.g. high-performance liquid chromatography (HPLC)), spectroscopic (e.g. MS), electrophoretic (e.g. two-dimensional sodium dodecyl sulfate-polyacrylamide gel electrophoresis (2D SDS-PAGE)), immuno-chemical (e.g. enzyme-linked immunosorbent assay (ELISA)) and NMR technologies, precede a longer rigorous process ultimately involving preclinical and clinical testing. However, it is important to note that very many drugs can fall along the expensive drug development path for any one of a number of reasons, and may even make it to market before adverse events are recognized. One example was the popular multi-billion dollar drug rofecoxib (*Vioxx*, cyclo-oxygenase-2 inhibitor), which despite proven efficacy was voluntarily withdrawn from the market due to adverse cardiovascular effects in some users. With the emergence of 'omics' comes the opportunity to couple the latest knowledge and understanding with advanced bioanalytical systems, to enhance the drug discovery process and move towards personalized medicine.

Drug design

The process of drug design is based on research, knowledge and understanding of potential, new, and established targets which historically have been proteins, primary examples of which are the so-called *G-protein coupled receptors* or *protein kinases*. Finding a new drug against a target is often easier said than done and is akin to finding a key that will fit and open a lock; while the full key may be known (e.g. peptide GLP-1 (glucagon-like peptide-1)) for a given lock (e.g. GLP-1 receptor), it is also possible that other keys (e.g. small molecules–natural or synthetic) can also fit into the lock, thereby triggering activity. Small molecules have often been considered preferential with respect to drug delivery by oral as opposed to injection route, but importantly may be less specific in action (i.e. trigger more than one receptor). Once a target has been established then a process of *high-throughput screening* (HTS) is begun, where many thousands or hundreds of thousands of chemicals can be tested for biological activity. Good examples are trying to find chemicals that can: (i) stimulate (agonist) or inhibit (antagonist) G-protein coupled receptors; or (ii) inhibit protein kinases.

Inherent in drug design is knowledge of biological activity, usually measured by quantitative functional assays (e.g. of target receptor interaction/activity/outcome); however, clearly it is also important to know the possible or actual toxicity of biologically active drugs. Knowledge of structure–activity relationships is important for drug discovery and *quantitative structure–activity relationship* (QSAR) modelling allows identification of chemical structures with high specificity, enhanced activity and low toxicity. QSAR techniques are versatile, which can also facilitate in-depth study of structural domains and active sites of biomolecules, and interactions can be quantitatively measured in natural or mutated targets (*site-directed mutagenesis*). While QSAR testing can go some way to reducing the use of animal testing, it is not a magic bullet and, as such, it would appear unlikely that QSAR testing will ever completely fulfil data requirements of regulatory bodies.

Drug testing

Importance of selectivity of a drug for a target cannot be understated, and of fundamental importance in drug testing is to identify the best molecule (*lead candidate*) for a given target. As indicated above, ideally the lead candidate, that must have current *good manufacturing practice* (cGMP) compliance, will not interfere with other targets (i.e. key will not activate other locks) or biological pathways, and often a *second candidate* will also be identified as a *backup*. Determination of *off-target toxicity* is vital before a lead candidate can proceed through drug development, and in particular clinical testing and knowledge of QSARs should improve the 'drug-like' or *pharmacokinetic* properties of the lead candidate. *Pharmacokinetics* (PK) is a branch of pharmacology that determines the fate of external agents (usually drugs) in living organisms, encompassing **A**bsorption (drug entering body), **D**istribution (dispersion or dissemination of drug through body), **M**etabolism (breakdown or transformation of drug), and **E**xcretion (elimination of drug or metabolites from body) – so-called **ADME**. *Pharmacodynamics* is a parallel field of study that examines what a drug does to the body, as opposed to PK that examines what the body does to the drug. ADME testing is one key aspect of preclinical drug testing which utilizes several animal models to pre-empt outcomes and give confidence of the effects of a drug before initiation of human clinical trials. The success is often associated with knowledge and understanding of drug targets and, given the expense of the drug development process, it is inherently more risky to develop drugs against new targets. Sometimes the potential or real toxicity of a drug is considered (ADME-Tox or ADMET), and ADME-Tox may be predicted by QSAR and other approaches for different routes of drug administration. Another important aspect of R&D of *new chemical enti-ties* (NCEs) relates to drug stability testing and related quality assurance in drug manufacture, storage and shelf-life, to further ensure maintenance of standards, patient safety, and satisfy key regulatory requirements.

Clinical drug development

After extensive preclinical *in vitro* and *in vivo* testing including toxicity testing in at least two separate animal species (e.g. rat and dog), applications to regulatory authorities (e.g. US Food and Drug Administration–FDA; European Medicines Agency–EMEA) for the use of NCEs in man are made, before commencing the first human clinical (so-called *first-in-man*) trials. Clinical drug development is closely monitored by drug licensing authorities that will be interested in the effects of the drug in healthy humans as well as intended patients (see Figure 12.2). Potential *long-term* and *multiple organ toxicity* and indeed *carcinogenicity* are always a concern, as well as more obvious *lack of efficacy* in humans, in which case the backup candidate may come into play. Pharmaceutical drugs typically undergo four key distinct clinical phases of drug development, termed *clini-cal trials*. After successful preclinical studies, appropriate regulatory approval

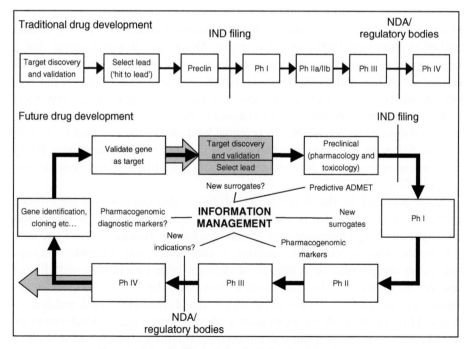

Figure 12.2   **Outline of the drug development process.**

(e.g. US FDA New Drug Application or NDA), and with sufficient funding in place, the decision to move to the clinic is made and *Phase I trials* begin using a small group (e.g. 50) of healthy volunteers to assess safety, tolerability, PK and pharmacodynamics of the drug in man. Initial safety having been established in Phase I facilitates the move to *Phase II trials* in a larger group (e.g. 200) to establish efficacy of the drug, where *Phase IIA* assesses dosing requirements and *Phase IIB* examines how well a given dose of drug works. *Phase III trials* are complex, expensive and time-consuming multicentre studies on larger patient groups (e.g. 2000), and assess how effective the drug is compared with current 'gold-standard' drugs. After Phase III trials all data are collected and prepared for submission and close scrutiny by regulatory authorities, before licensing and entry of the drug to the market. *Phase IV trials* monitor safety of drugs after they go on sale, identifying possible rare or long-term adverse effects in the population, which may result in refining the use of a drug to particular population groups, or in some instances withdrawal. In the future, the process of clinical drug development will be informed by availability of 'omic' information and markers that can be used to develop better pharmaceutical products and realize the ultimate goal of personalized medicines.

12.4   Emerging pharmaceutical products

Historically, drug discovery and development has been on the basis of only around 500 of an estimated 1500 *druggable* targets. Given this, it would appear that a large number of new drugs exist that can be *mined* from various natural sources (including NCEs), and using emerging virtual screening technologies a pipeline of compounds can be established for testing. Clearly the drug discovery process is under constant analysis and scrutiny to increase the efficiency of identifying 'hits' (potential drugs) and 'leads' (best of potential hits), during the so-called *hit-to-lead* process. Scientists will often integrate information from multiple technology-based strategies to optimize this process, and a few of these strategies and examples are given in Table 12.2.

Much potential in the development of future therapeutics will come from *translational research*, that is, the collaboration and partnership of scientists and clinicians allowing implementation of complementary strategies, where one area informs the other. Indeed, with electronic health records and electronic data capture come the promise of collection of data for both patient care and clinical research databases. Translational research allows capitalization on the latest breakthroughs in the 'omics' to best understand functional changes in health and disease states. This, in turn, allows identification of new drug targets, development of *linked animal models* and clinical biomarkers to enhance preclinical and clinical study design and the drug development process. Natural products have long

Table 12.2   **Examples of technology-based strategies to optimize 'hit-to-lead' process**

Strategy	Brief description	Example(s)
High-throughput biochemical and cellular assays	Hit identification by HTS of large numbers of small molecules	Identification of compounds against adipocyte fatty acid binding protein (FABP4)
Assay of natural products	Isolation of active ingredient(s) from natural sources for evolution of new drug classes	Generation of 3-hydroxy-3-methyl-glutaryl-coenzyme A (HMG-CoA) reductase inhibitors from isolated active biomolecules from microbes
Structure-based design	Structural analyses and modelling of receptor–ligand interactions	Generation of protein phosphatise 1B inhibitors from structural data
Peptides and peptidomimetics	Definition of critical minimum sequence structure–activity relationships for target modulation and isolation, identification, design and synthesis of peptide analogues	Isolation of GLP-1 receptor agonist, exendin, from lizard venom

been believed to be rich sources of chemical entities with medicinal potential. For example, venoms and skin secretions of amphibians are known to contain rich cocktails of molecules, which may poison their prey or offer protection against predators. Protection may be afforded against other threats such as infection, and considering some amphibians live happily in swamps, some of the skin secretions may contain agents offering antibacterial action. Such molecules could represent important new antibacterial drugs and it is interesting to note that of the NCEs isolated from natural products, by far the most prevalent medicinal indications are antibacterial and anticancer. However, a good recent example of translational research in the field of natural products is the discovery and exploitation of exendin, a naturally occurring peptide that has been developed and marketed (as exenatide, *Byetta*) for the treatment of Type 2 diabetes, as outlined in Chapter 1. Other classic examples of pharmaceutical products derived from natural sources are outlined in Table 12.3.

Much excitement surrounds the rise of *biologics* as therapeutic products. 'Biologics' is an umbrella term covering a wide spectrum of potential medicinal products including vaccines, blood/blood components, recombinant proteins, somatic cells and gene therapy. As the name suggests, biologics are isolated from a number of natural sources (animal, human, or micro-organism) and may be generated by

Table 12.3   **Examples of emerging pharmaceutical products**

Source	Agent	Medical indication
Plant	Digitalis	Cardiovascular (antiarrhythmic)
	Morphine	Pain (opiate analgesic)
Microbe	Penicillin	Bacterial infections (β-lactam antibiotic)
	Lovastatin	High cholesterol (hypolipidaemic statin)
Animal (frog) (snake)	Epibatidine	Pain (non-opiate analgesic)
	Teprotide	Cardiovascular (antihypertensive)

biotechnology and other state-of-the-art approaches. Nonliving biologics composed of sugars/proteins/nucleic acids or complex combinations may offer useful alternatives to 'small molecule drugs', but with often higher molecular complexity comes related manufacturing challenges. Cell- and gene-based therapeutic biologics are particularly exciting approaches to conditions for which there are no other treatments or which rely on strict compliance with intensive drug application. Examples of biologics emerging from recombinant DNA technology include: (i) stimulator of red blood cell production – erythropoietin (*Epogen*, epoetin alfa), a recombinant protein; (ii) tumour necrosis factor (TNF) antagonist rheumatoid arthritis drugs – adalimumab (*Humira*), a monoclonal antibody, and etanercept (*Enbrel*), a recombinant human TNF receptor fusion protein; and (iii) human epidermal growth factor (HER2)/neu (erbB2) antagonist breast cancer agent – trastuzumab (*Herceptin*), a humanized monoclonal antibody.

12.5   Future perspectives

There continue to be rapid technological advances in the field of bioanalysis which are focused on making the processes of R&D in diagnostics and drug discovery more successful and efficient, with the aspiration to make processes and products less labour-intensive and more cost-effective. The following subsections provide some highlights and indications of up-and-coming approaches which may be 'state-of-the-art' today but 'matter-of-fact' tomorrow.

'Lab-on-a-chip' (LOC): The LOC concept is based on extreme miniaturization of laboratory tests or functions on a single platform (chip) of millimetre or centimetre scale. LOC devices being developed allow very small volumes of sample (picolitres) to be loaded onto the chip for analysis, where multiple tests can take place. The first reported LOC device was a gas chromatograph in the 1970s, however it wasn't until the 1990s that the promise of this technology started to be realized. While the immediate future of LOC technologies includes real-time polymerase

chain reaction (PCR), immunoassay and cellular ion channel screening, advances in nanotechnology offer much promise with applications in analytical (e.g. medical diagnostics, forensics and environmental monitoring) and pharmaceutical (e.g. rapid HTS and microreactor) fields.

Biochip and microarray technologies: Advances in biochip technologies mirror R&D efforts in a number of emerging fields including genomics, proteomics, pharmaceuticals, and systems biology. Biochips fuse biotechnological processes with sensor technologies into miniature chip devices, offering the opportunity to conduct hundreds or even thousands of bioanalytical reactions simultaneously. One of the first commercial uses of biochip technology was by the company Affymetrix, in their so-called *GeneChip* products, containing grids of thousands of individual *'DNA sensors'* to detect genes such as *BRCA1* and *BRCA2* (related to breast cancer) through incorporation of so-called *microarray technology* and *surface chemistry*. However, microarrays are not limited to analysis of DNA and protein; antibody and chemical microarrays are also being produced as biochips, where the Northern Ireland-based company, *Randox Laboratories*, introduced the first protein biochip array analyser (*Evidence*). The Evidence device launched in 2003, and can be used in patient profiling for disease screening, diagnosis and monitoring of disease progression/treatment.

'Point-of-care testing' (POCT): This represents technologies which are available at the site of patient care, typically for adoption in a 'doctor's surgery', at the 'bedside' or even in the 'home'. Ideally POCT relies on small, portable and easy-to-operate instruments. A simple and good example of POCT is the blood glucose meter used for diabetes monitoring/testing (see Chapter 4). With increasing miniaturization, portability and reduced device/consumable cost, POCT is diversifying far outside the traditional hospital environment, playing key roles in public and environmental health and other emerging applications such as forensics and national defence against bioterrorism.

Computer-aided drug design: In the postgenomic era the evolution and increasing use of computer-aided drug design (CADD) offers considerable potential in most aspects of drug discovery, from target identification and lead discovery through to clinical trials. Indeed, CADD methods can be used for NCE discovery through *de novo* drug design (building a complete molecule from 'molecular bricks'), *combinatorial chemistry* and *library design* (allows thousands/millions of compounds to be synthesized simultaneously), virtual screening, drug-like analysis and ADME-Tox prediction. As CADD develops to integrate *computational chemistry* and biology together with *chemoinformatics* and bioinformatics (so-called *pharmacoinformatics*), it will no doubt impact on the pharmaceutical development process and increase the success rate of drug candidates.

Development of specific drugs by targeted structural genomics: Active site similarities of enzymes such as protein kinases can lead to lack of selectivity and unwanted clinical side effects. Structural genomics facilitates determination of the human protein structures, and developments in biotechnology and protein crystallography have already yielded valuable data for structure-guided drug discovery. The human kinome represents >500 protein kinases grouped into 10 families representing some 1.7% of all genes. The use of structural genomics should help identify novel structural features (motifs), knowledge of which should help develop more specific protein kinase inhibitors.

Pharmacogenetics in drug development: With better understanding of the molecular basis of disease comes the promise of developing improved medicines. Genetic association studies can help define pathways linked with disease processes, and pharmacogenetics can assist in streamlining of drug development whereby variability in drug response may be explained by genetic polymorphisms of drug targets (so-called *candidate gene linkage studies*). Furthermore, availability of DNA samples from Phase III/IV trials may further extend novel target identification, enabling so-called *whole-genome association studies*. The potential of pharmacogenetics in the development of personalized medicines, where the drug is targeted in persons most likely to respond, is highlighted by the development of the HER2 monoclonal antibody trastuzumab mentioned earlier, which is highly effective in women overexpressing the HER2/neu receptor, representing ~30% of breast cancer patients.

Personalized medicine: Personalized drug-discovery efforts offer the promise of great progress in major therapeutic areas, offering treatment best suited for a given individual based on genomic and other key factors regulating drug responses. Molecular biomarkers represent an important link between drug discovery and diagnostics with the integration of emerging technologies (e.g. RNA interference, nanobiology and high-throughput genotyping) and increasing knowledge of human genetic variation. Personalized medicine offers the promise of molecular markers to identify risk of disease, before onset or exhibited clinical signs and symptoms, underlying new strategies focused on prevention and early intervention. The molecular basis of personalized medicine includes detection of gene sequences and characteristic gene, protein and metabolite profiles associated with disease states. This information should enable targeted screening and tailor-made therapeutic interventions in genetic subsets of patients which will best respond. Furthermore, this approach should help predict outcomes based on genetic information before drug administration and allow selection of drugs and doses suitable for individual genetic backgrounds. With parallel development of simple and clinically convenient genetic analyses, such tailor-made interventions should also avoid wasteful drug administration and associated economic costs. Given this, while pharmacogenetics predicts drug efficacy and toxicity, pharmacoeconomics aims to deliver more cost-effective healthcare. In addition to the HER2 expression/trastuzumab responsiveness noted earlier, other primary evidence comes from screening for: (i) thiopurine methyltransferase gene polymorphisms to prevent azathioprine-induced myelosuppression; (ii) human leukocyte antigen B5701 to prevent hypersensitivity to the drug abacavir; and (iii) angiotensin-converting enzyme polymorphisms for statin therapy.

Proteomics for cancer: 2D-PAGE and MS analyses have already generated large datasets of potential diagnostic, prognostic and therapeutic importance. Other advanced and emerging bioanalytical methods such as laser capture microdissection, isotope-coded affinity tag technology, reverse-phase protein arrays, surface-enhanced laser desorption/ionization time-of-flight (SELDI-TOF) analysis and protein microarrays, coupled with messenger ribonucleic acid (mRNA)-based assays and improved clinical data should help identify novel diagnostic biomarkers for cancer and other diseases. While the inherent challenges in therapeutic proteomics lie in the complexity of systems and cells which generate hundreds of thousands of different polypeptides, the synthesis and comparison of datasets from large numbers of normal and diseased individuals (*differential expression*) should help profile target pathways before, during and after therapy, and answer other key clinical questions in cancer research. Importantly, metabolomics also utilizes advanced approaches to molecular phenotype analysis and may offer new insights into biomarker patterns associated with health and disease, for personalized health monitoring and design of individual interventions.

New indications for established drugs: As well as reformulation of established drugs to enhance safety, efficacy and convenience ('new drugs from old'), translational research offers great potential in identifying new therapeutic uses of known drugs, so-called *indication discovery*. Examples of success to date include sidenafil (*Pfizer's Viagra*), a selective phosphodiesterase type 5 (PDE5) inhibitor originally used for treatment of angina, now popularly known as an effective oral treatment for male erectile dysfunction … a memorable if not interesting way to end this chapter, and the main body of this textbook!

Key Points

- Advanced bioanalytical tools provide the means of studying emerging 'omics' fields, which encompass the totality of clinical genomics, proteomics and metabolomics.

- A primary goal of clinical genomics, proteomics and metabolomics is the identification of early disease biomarkers and therapeutic drug targets.

- Molecular genetics including direct, linkage-based and genomic testing provides important diagnostic and predictive information for robust analysis of the genetic heritability of a disease.

- Four primary indications/applications of genetic testing are: diagnostic testing, pre-symptomatic testing, pre-dispositional (susceptibility) testing, and pharmacogenetic testing.

- Bioanalytical chemistry feeds R&D, taking it to another level of scale or complexity, with the aim of generating new drugs, diagnostic agents, technologies or other commercial products.

- The drug discovery process involves key steps from identification of lead candidate through drug design and testing to clinical drug development (Phase I–IV trials).

- Translational research represents an R&D interface where collaboration and partnership of scientists and clinicians facilitates understanding of functional changes in health and disease states.

- Biologics are widely hailed as the next generation of medicinal products encompassing vaccines, blood/ blood components, recombinant proteins, somatic cells and gene therapy.

- Rapid technological advances enable future R&D processes in diagnostics and drug discovery, realizing more specific, successful, efficient and cost-effective tools for bioanalysis.

- Novel approaches to bioanalysis and testing (e.g. LOC, biochip and POCT) coupled with advances in CADD, genomics and proteomics should help realize the ultimate goal of personalized medicine.

13 Self-assessment

Question 1:

A nucleotide has three key components which are:

(A) Phosphate group(s); a base; a sugar

(B) Sulfate group(s); an acid; a sugar

(C) Phosphate group(s); a base; an aldehyde

(D) Sulfate group(s); an acid; an aldehyde

Question 2:

The amino acid tryptophan is represented by the one letter _____ and like all other amino acids contains a _____ group and a _____ group.

(A) T; COOH; NH_2

(B) T; COOH; CH_3

(C) W; COOH; NH_2

(D) R; COOH; CH_3

Question 3:

In which of the following are all the biomolecules lipids?

(A) Cholesterol; ceramine; stearic acid

Understanding Bioanalytical Chemistry: Principles and applications   Victor A. Gault and Neville H. McClenaghan
© 2009 John Wiley & Sons, Ltd

(B) Stearic acid; glycerol; phosphatidyinositol

(C) Glyceraldehyde; cholesterol; phosphatidylinositol

(D) Stearic acid; ceramine; glycerol

Question 4:

How many moles are there in 6.5 g of glycine (RFM = 75 g)?

Question 5:

You have prepared 180 ml of a 2.8 M stock buffer solution, but you require 0.95 M. How many millilitres of the 0.95 M solution can you make?

Question 6:

Calculate the molarity when 36 g of NaCl (RFM = 58.5 g) is dissolved in 1250 ml of water.

Question 7:

Key reactions of transition metals include:

(A) Electron transfer; insertion; redox-catalysed deletion

(B) Substitution; deactivation of ligands; redox-catalysed insertion

(C) Redox-catalysed substitution; proton transfer; activation of ligands

(D) Substitution; oxidative-addition reaction; redox-catalysed insertion

Question 8:

Which of the following disorders is associated with defective transport of transition metals?

(A) Menkes' syndrome

(B) Wilson's disease

(C) Haemochromatosis

(D) Reynold's syndrome

Question 9:

Which of the following can comprise the reference electrode in a pH measuring device?

(A) Sodium chloride

(B) Mercury chloride

(C) Potassium sulfate

(D) Hydrogen electrode

Question 10:

If the pH of a solution is 6.32, what is the hydrogen ion concentration?

Question 11:

Calculate the pH of a solution with a hydrogen ion concentration of 3.4×10^{-6} M.

Question 12:

A biomolecule has an extinction coefficient of $15000\,M^{-1}\,cm^{-1}$ at 260 nm in a 1 cm pathlength cuvette. From this information calculate the concentration of this biomolecule in a solution which gave an absorbance reading of 0.60.

(A) 9 mM

(B) 3.9 mM

(C) 40 µM

(D) 25 µM

Question 13:

In infrared (IR) spectroscopy, the fingerprint region is defined as:

(A) $1900-1500\,cm^{-1}$

(B) $2500-1900\,cm^{-1}$

(C) $<1500\,cm^{-1}$

(D) $4000-2500\,cm^{-1}$

Question 14:

Which types of electronic transitions are associated with ultraviolet/visible spectroscopy?

Question 15:

Which of the following factors will not affect the rate of sedimentation of a particle during centrifugation?

(A) Size

(B) Shape

(C) Density

(D) Atmospheric pressure

Question 16:

Which of the following is not a medium used for density gradient centrifugation?

(A) caesium chloride

(B) dextran

(C) sucrase

(D) methyl glucamine salt of triiodobenzoic acid

Question 17:

Low-speed centrifuges can be used to pellet what?

(A) Ribosomes

(B) Cell organelles

(C) Macromolecules

(D) Nuclei

Question 18:

During paper chromatography a mixture of biomolecules was added to the stationary phase at the origin and several hours after addition of the mobile phase a number of spots were visible at distances of (a) 5 cm, (b) 10 cm, and (c) 15 cm from the origin. During this time the mobile phase moved a total distance of 20 cm (the solvent front). Using this information calculate the R_f values of each of the three separated biomolecules.

Question 19:

Examples of popular chemical reagents used in chromatography for visualization and detection are _____ (for amino acids), _____ (for lipids) and _____ (for carbohydrates).

Question 20:

An *anion exchanger* comprises a resin containing _____ charged (_____) functional groups, while the resin in a *cation exchanger* contains _____ charged (_____) functional groups.

Question 21:

List the four key components of an HPLC system.

Question 22:

Which of the following is a key physical characteristic underlying electrophoretic separation?

(A) Net charge

(B) Colour

(C) Weight

(D) Density

Question 23:

Agarose gel electrophoresis is used primarily in the separation of:

(A) Proteins

(B) Peptides

(C) DNA

(D) Fragments of nucleic acids

Question 24:

Which of the following is not used during staining after polyacrylamide gel electrophoresis?

(A) Coomassie Brilliant Blue

(B) Amido black

(C) Silver nitrate

(D) Ponceau blue

Question 25:

List four advanced electrophoretic separation methodologies for genomic and proteomic analyses.

Question 26:

Which of the following best describes the general configuration of any mass spectrometer?

(A) Ion source; MALDI analyser; detector

(B) Syringe pump; ion-trap analyser; detector

(C) Ion source; mass analyser; detector

(D) Syringe pump; MALDI analyser; detector

Question 27:

Which of the following is not a MS analyser?

(A) Quadrolite

(B) Time-of-flight

(C) Ion-trap

(D) Orbitrap

Question 28:

Which of the following is the least important property of an antibody when used in an immunoassay?

(A) Range

(B) Specificity

(C) Affinity

(D) Size

Question 29:

In a radioimmunoassay, dextran-coated charcoal is primarily used to:

(A) Purify antisera

(B) Separate free antigen from antibody–antigen complexes

(C) Prevent non-specific binding

(D) Purify radiolabelled tracer

Question 30:

In an indirect ELISA:

(A) The plate is coated with antibody

(B) The antigen is sandwiched between antibody on the plate and the labelled antibody

(C) The labelled antibody binds directly to the antigen coating the plate

(D) The labelled antibody binds to a primary antibody attached to the antigen coating the plate

Question 31:

Which of the following is not an essential requirement for NMR recordings?

(A) Contrast agent

(B) Intense static magnetic field

(C) High frequency transmitter

(D) Detector

Question 32:

MRI relies solely on the _____ properties of the _____ atom to produce images (comparable to _____), representing 'non-invasive surgical examination'.

Question 33:

List three major diagnostic applications of MRI in specialized scans.

Question 34:

A primary goal of clinical genomics, proteomics and metabolomics is the identification of early _____ and therapeutic _____ .

Question 35:

List four primary indication/applications of genetic testing.

Appendix 1: International system of units (SI) and common SI prefixes

International system of units (SI)

Quantity	Name	Abbreviation/symbol
Length	metre	m
Time	second	s
Mass	kilogram	kg
Temperature	kelvin	K
Volume	litre	L
Amount of substance	mole	mol
Electric current	ampere	A
Radioactivity	becquerel	Bq

Common SI prefixes

Name	Abbreviation/symbol	Multiplication factor
giga	G	10^9
mega	M	10^6
kilo	k	10^3
deca	da	10
milli	m	10^{-3}
micro	μ	10^{-6}
nano	n	10^{-9}
pico	p	10^{-12}
femto	f	10^{-15}

Understanding Bioanalytical Chemistry: Principles and applications   Victor A. Gault and Neville H. McClenaghan
© 2009 John Wiley & Sons, Ltd

Appendix 2: The periodic table of the elements

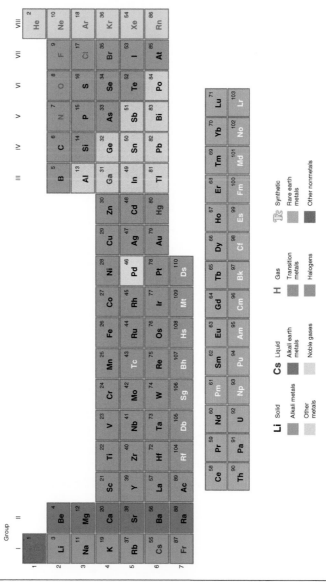

Understanding Bioanalytical Chemistry: Principles and applications   Victor A. Gault and Neville H. McClenaghan
© 2009 John Wiley & Sons, Ltd

Appendix 3: Common solvents and biological buffers

Common solvents

Name	Boiling point (°C)
Acetic acid	118
Acetone	56
Acetonitrile	82
Benzene	80
n-Butanol	118
Chloroform	61
Dichloromethane	40
Diethyl ether	35
Dimethyl formamide	153
Dimethyl sulfoxide	189
Ethanol	79
Ethyl acetate	77
Formic acid	100
Hexane	69
Methanol	65
n-Propanol	97
Tetrahydrofuran	66
Toluene	111
Water	100

Understanding Bioanalytical Chemistry: Principles and applications   Victor A. Gault and Neville H. McClenaghan
© 2009 John Wiley & Sons, Ltd

Common biological buffers

Name	Boiling point (°C)
Acetate	3.8–5.8
Bicine	7.8–8.8
HEPES	7.0–8.0
HEPPS	7.6–8.6
MOPS	6.5–7.9
TAPS	7.7–9.1
TES	6.8–8.2
Tricine	7.4–8.8
Tris	7.5–9.0

Appendix 4: Answers to self-assessment questions

Question 1: A

Question 2: C

Question 3: B

Question 4: 0.867 mol

Question 5: 530.5 ml

Question 6: 0.49 M

Question 7: D

Question 8: A

Question 9: B

Question 10: 4.79×10^{-7} M

Question 11: 4.47

Question 12: C

Question 13: C

Question 14: n $\rightarrow \pi^*$ and $\pi \rightarrow \pi^*$ transitions

Question 15: D

Question 16: C

Question 17: D

Question 18: (a) 0.25, (b) 0.5 and (c) 0.75

Question 19: Examples of popular chemical reagents used in chromatography for visualization and detection are **ninhydrin** (for amino acids), **rhodamine B** (for lipids) and **aniline phthalate** (for carbohydrates).

Understanding Bioanalytical Chemistry: Principles and applications   Victor A. Gault and Neville H. McClenaghan
© 2009 John Wiley & Sons, Ltd

Question 20: An *anion exchanger* comprises a resin containing **positively** charged (**basic**) functional groups while the resin in a *cation exchanger* contains **negatively** charged (**acidic**) functional groups.

Question 21: Solvent pump (delivery of mobile phase); sample injector; HPLC column (stationary phase); and detector linked to recording device.

Question 22: A

Question 23: C

Question 24: D

Question 25: PCR; Isoelectric focusing; two-dimensional gel electrophoresis; capillary gel electrophoresis and pulsed-field gel electrophoresis

Question 26: C

Question 27: A

Question 28: D

Question 29: B

Question 30: D

Question 31: A

Question 32: MRI relies solely on the **magnetic** properties of the **hydrogen** atom to produce images (comparable to ^{1}**H NMR**), representing 'non-invasive surgical examination'.

Question 33: Any three from: MRA; diffusion MRI; perfusion MRI; fMRI; and MRS

Question 34: A primary goal of clinical genomics, proteomics and metabolomics is the identification of early **disease biomarkers** and therapeutic **drug targets**.

Question 35: Diagnostic testing; pre-symptomatic testing; pre-dispositional (susceptibility) testing; and pharmacogenetic testing.

INDEX

References in *italics* are to pages with figures, references in **bold** are to pages with tables

Understanding Bioanalytical Chemistry: Principles and applications Victor A. Gault and Neville H. McClenaghan
© 2009 John Wiley & Sons, Ltd

Printed and bound by CPI Group (UK) Ltd, Croydon, CR0 4YY

27/10/2024

14580195-0002